Intelligent Building
Dictionary

terminology for smart, integrated, green building design, construction, and management

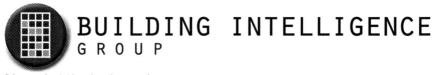 BUILDING INTELLIGENCE
GROUP

Chuck Ehrlich, editor

Hands-on-Guide
San Francisco

Intelligent Building Dictionary: terminology for smart, integrated, green building design, construction, and management
by Building Intelligence Group LLC, edited by Chuck Ehrlich
© 2007 by Hands-on-Guide

Hands-on-Guide
2929 Webster St.
San Francisco, CA 94123 USA
www.HandsOnGuide.com   sales@handsonguide.com

ISBN 10: 0-9796408-4-9   ISBN 13: 978-0-9796408-4-1
Library of Congress Control Number: 2007904424
Publishing history: first edition, revision 1.01

This book is available in bulk at quantity discounts; please e-mail sales@handsonguide.com for more information.

Introduction

Intelligent Building technology provides important benefits including:

- Energy efficiency and sustainability: buildings account for half of CO_2 emissions contributing to global warming.

- Health, comfort and safety of building occupants.

- Productivity and efficiency of occupants and owners.

Designing, building, and operating an Intelligent Building is a team effort that involves an exceptionally wide range of disciplines as shown in Table 1 on the next page. What distinguishes Intelligent Buildings is the combination of:

- Green Building focus together with integration and optimization of all building systems

- Integrated and more sophisticated control systems for HVAC/personal comfort, lighting, active facades, daylight control, audiovisual equipment, sub-metering, energy management, digital signage, etc.

- Information Technology in a central role to integrate building systems with other building systems and with enterprise information systems.

This convergence of disciplines can result in confusion since each group has their own vocabulary and groups sometimes use the same term with different meanings. This dictionary is intended to facilitate cooperation by providing a common reference for all team members that includes acronyms, technical terms, and jargon.

Technology and terminology will continues to evolve, and we plan to revise this dictionary to keep up with these changes. Please help us with this process by going to www.Intelligent-Building-Dictionary.com and letting us know what terms we missed, what definitions are not clear, and if you disagree with our definition. Thank you.

Chuck Ehrlich, editor

Table 1: Intelligent Building Roles	
Acoustic Engineer	Information Technology
Alternative Energy Specialist	Interface Designer
Architect	Interior Design
Audiovisual Designer	Landscape Architect
Backup Power Engineer	LEED Accredited Professional
Building Owners	Lighting Design Engineer
Carpenter	Lighting Designer
Chief Facilities Officer	Mechanical Engineer
Chief Information Officer	MEP Engineer
Chief Security Officer	Network Engineer
Civil Engineer	Network Manager
Commissioning Agent	Operations Manager
Communication, Life-safety, and	Plumber
Automation Consultant	Process Engineer
Construction Manager	Product Representative
Construction Specifier	Professional Engineer
Consulting Engineer	Program Manager
Contract Administrator	Programmer
Contractor	Project Manager
Control Systems Engineer	Risk Manager
Data Center Designer	Security Specialist
Daylighting Designer	Space Planner
Developer	Structural Engineer
Documentation Specialist	Sustainable Design Engineer
Electrician	Systems Administrator
Energy Engineer	Systems Integrator
Energy Manager	Telecommunications Specialist
Facilities Engineer	Transport Engineer
Facilities Manager	User Interface Designer
HVAC Designer	Wireless/Radio Engineer
Indoor Air Quality Professional	Workflow Designer

Acknowledgements

The authors and editor would like to thank our reviewers, especially:

- Kitty Myers, AIA
- Jeff Seewald

Their time and attention to detail has helped make this a better book.

We would also like to thank our families for their support.

Building Intelligence Group would like to thank our clients, without whom none of this would be possible.

Table of Contents

Table of Figures

Explanatory Notes

Words and Figures

This dictionary has two parts, the Words section contains the definitions of terms and the Figures section contains tables and diagrams. Figures are cross referenced by number and page number.

Guide Words

A pair of guide words is printed at the top of every page as an aid in locating entries. The word to the left of the bullet is the first word on the page and the word to the right of the bullet is the last entry on the page.

Entry Order

Entry words are printed in bold type and positioned slightly to the left of the text column. All entries have been listed in alphabetical order. Terms the begin with punctuation or numbers appear before letters.

Bold Small Caps

Boldface and small caps indicate a TERM that is defined elsewhere in this dictionary.

Acronyms

Acronyms are shown without periods and with their full form. Acronyms with only one meaning are shown with their full form and the definition of the full form as a convenience for the reader.

Words

.NET or **Dot Net,** a Microsoft framework for distributed software applications that are platform independent and may use multiple programming languages.

1000Base-T or **802.3ab,** ETHERNET at 1000 Mb/s or 1Gb/s over Cat 5e cable.

100Base-T 802.3u, or **802.3y,** ETHERNET at 100 Mb/s.

1080i a HIGH DEFINITION TELEVISION video mode with 1080 lines of vertical resolution, interlaced or non-progressive scan. Field rate may be implied or appended as in 1080i50 for 50 Hz or 1080i60 for 60 Hz; frame rate (half of the field rate) may follow with a slash as in 1080i/25 or 1080i/30.

1080p a HIGH DEFINITION TELEVISION video mode with 1080 lines of vertical resolution, progressive scan or non-interlaced. Field rate may be implied or appended as in 1080p50 for 50 Hz or 1080p60 for 60 Hz; frame rate (half of the field rate) may follow with a slash as in 1080p/25 or 1080p/30.

10Base-T or **802.3i,** Ethernet at 10 Mb/s over Cat 3 cable.

10GBase-T or **802.3an,** Ethernet at 10 Gb/s over Cat 6a cable.

1x 1xRTT, or **IS-2000,** 1 times Radio Transmission Technology, a mobile data service available on **CDMA2000** cellular telephone networks.

24/7 or **24x7**, shorthand for 24 hours a day and seven days per week; in other words the CONTINUOUS AVAILABILITY of a service or resource.

3D CAD a three dimensional COMPUTER AIDED DESIGN model or modeling software.

3DES Triple **DES**, a popular but nonstandard abbreviation for the TRIPLE DATA ENCRYPTION ALGORITHM.

3G third-generation technology for mobile telephones capable of supporting a variety of services including voice telephony, broadband wireless data, and video. See Fig. 9 on page 285 and Fig. 10 on on page 286.

3GPP2 Third Generation Partnership Project 2, a collaborative effort between North American and Asian standards development organizations covering high speed, broadband, and Internet Protocol-based mobile systems. For more information see www.3gpp2.org.

3GPP LTE Third Generation Partnership Project Long Term Evolution, a project to enhance **UMTS** as a fourth-generation network based on a wireless broadband IP network supporting voice and other services.

3PL Third Party Logistics, an organization that manages the movement of goods for other companies.

480p video display resolution with 480 vertical lines and progressive (non-interlaced) scan.

4-PCH or **4-Phenylcyclohexene**, a chemical by-product formed during the manufacture of certain types of carpet backing. Exposure to 4-PCH or other emission products following new carpet installation may result in health complaints.

4D Model a software tool for visualizing the three dimensional model (3D CAD) of a project together with the construction schedule in order to analyze changes over time so that work schedules may be coordinated and optimized.

5 Nines 99.999% AVAILABILITY for a system or service. See Fig. 4 on page 281.

6LoWPAN Wireless Personal Area Network for IPv6, standard under development for **IPv6** over IEEE 802.15.4 networks. See www.6lowpan.net for more information.

720p video display resolution with 720 vertical lines and progressive (non-interlaced) scan.

802.11 or IEEE 802.11, a family of standards for wireless Ethernet networks, commonly known as WI-FI. 802.11b provides a raw data rate of 11 Mbit/s and 802.11g provides 54 Mbit/s using the 2.4 gigahertz (GHz) radio band, 802.11a provides 54 Mbit/s using the 5 GHz band.

802.11i or IEEE 802.11i, an amendment to the IEEE 802.11 standard specifying additional security mechanisms for WI-FI wireless networks.

802.15.1 or IEEE 802.15.1, the standard for BLUETOOTH wireless networks providing a data rate of up to 3 Mbit/s over short distances (1, 10, or 100 meters depending on the device class) using the 2.45 GHz radio band.

802.15.4 or IEEE 802.15.4, the standard for ZIGBEE wireless networks designed for low data rate, long battery life, and secure networking using ISM BAND frequencies: 868 MHz in Europe, 915 MHz in the USA and 2.4 GHz in most jurisdictions.

802.16 or IEEE 802.16 WirelessMAN, the standard for WIMAX wireless networks providing high speeds up to 70 Mbit/s over short distances and lower speeds over longer distances (10 Mbit/s over 10 km) using a variety of radio frequencies (2.3, 2.5, 3.5, 3.7 and 5.8 GHz).

802.1D or IEEE 802.1D, standard for Local and Metropolitan Area Networks—MEDIA ACCESS CONTROL (MAC) Bridges technology for connecting together LANs at different locations in a transparent fashion using MAC Bridges. The 2004 update to this standard incorporates

RAPID SPANNING TREE PROTOCOL and obsoletes SPANNING TREE PROTOCOL for routing bridged LAN networks.

802.1Q or **IEEE 802.1Q,** standard for a Virtual Bridged Local Area Network or **VLAN.**

802.1X or **IEEE 802.1X,** Port Based Network Access Control standard authenticates devices attached to a LAN port using **EAP.**

802.3 IEEE 802.3 or **ISO 8802,** family of standards for wired Ethernet networking using various types of copper or fiber optic cables.

802.3af or **IEEE 802.3af,** the standard for POWER OVER ETHERNET.

A

A&E Architects and Engineers or architecture and engineering firms.

A/D Analog-to-Digital.

A/E/C Architects/Engineers/Contractors.

A/V Audio/Video or Audio/Visual, see AUDIOVISUAL.

AAA (Authentication, Authorization, and Accounting) can be either a server providing these functions or the protocol for accessing and sharing this information as defined in RFC-2903 and related documents.

AAC ADVANCED APPLICATION CONTROLLER or ADVANCED AUDIO CODING.

AACS (Advanced Access Content System), the content protection scheme used for video on DVD.

ABC (Activity Based Costing) an accounting technique for associating costs with each business process.

Absorption the physical process where energy is transfered from one particle to another; for example when a photon of light falls on an object and makes it warmer. For example, HVAC uses absorption to transfer heat energy from the medium being cooled to the refrigerant.

Absorption Chiller cooling technology based on a thermal or chemical process instead of mechanically compressing the refrigerant as in traditional vapor compression chillers. Most absorption chillers utilize lithium bromide (a salt) and water as the fluid pair.

Abstraction Layer or **Abstraction Level,** a technique for hiding details or encapsulating a group functions by defining an idealized interface between layers or levels. For example, see the **OSI MODEL.**

ACATS Advisory Committee on Advanced Television Service, a group established to give the **FCC** advice on issues related to advanced television services.

Access Control the process of restricting entrance to physical space, communications network or system including processes to AUTHENTI-CATE permitted users, CREDENTIAL users and grant appropriate access rights or privileges, and maintain records of access and activities. See also **IDENTIFICATION AND AUTHENTICATION.**

Access Control List information that is used to restrict access to authorized users or groups of users. See also **WHITE LIST** and **BLACKLIST.**

Access Control System business processes and technology that support the ACCESS CONTROL process and maintain the necessary records. Physical access controls may include humans, mechanical barriers, key locks, card locks, combination locks, etc.

Access Point a device that connects devices or networks together.

ACD (Automatic Call Distributor) a **PBX** or other telephone switching system with special features for handling large volumes of incoming calls and queuing them for multiple agent groups or other resources.

aceXML a group within **IAI** formed to facilitate interoperability between software through the use of XML technology in architecture, engineering, and construction including design, construction, and life cycle applications. See www.iai-na.org/aecxml for more information.

ACH Air Change rate per Hour, a measure of FUME HOOD effectiveness.

ACID Atomicity, Consistency, Isolation, Durability, properties of a database system that assure reliable TRANSACTION processing: atomicity implies all or nothing, see ATOMIC TRANSACTION, consistency refers to integrity constraints on database content, isolation implies hiding partially processed transactions, and durability implies RELIABILITY and the ability to recover from certain types of failures without losing or duplicating transactions.

ACK (Acknowledge) or **Acknowledgement,** to confirm that a message, signal, or alarm has been received or recognized. A user may acknowledge a message by pressing a button or issuing a software command, a system may send an acknowledgement character or protocol message.

Acknowledge or **Acknowledgement,** to confirm that a message, signal, or alarm has been received or recognized. A user may acknowledge a message by pressing a button or issuing a software command, a system may send an acknowledgement character or protocol message.

ACL (Access Control List) information that is used to restrict access to authorized users or groups of users. See also WHITE LIST and BLACKLIST.

Active Directory Microsoft's **LDAP** directory services implementation for the Windows environment.

Active Façade a building exterior equipped with SMART GLASS, shading systems, or other technologies that can dynamically change the optical and thermal transmission characteristics of the windows.

Active IR signaling using infrared light from a powered device such as a remote control. Contrast to PASSIVE IR.

Active RFID use of **RFID** tags that contain an internal power source and can report information automatically. Contrast to **Passive RFID**.

Active Server Pages a Microsoft facility for dynamically generating Web pages using scripts or programs that run in the **server**.

Active Solar a general class of technologies that convert solar energy into usable heat, cause air-movement for ventilation or cooling, or store heat for future use, and use electrical or mechanical equipment, such as pumps and fans, for increased efficiency. Contrast to **Passive Solar**.

ActiveX Control a reusable software component based on Microsoft's version of the Component Object Model (**COM**).

Activity Based Costing an accounting technique for associating costs with each business process.

Ad-hoc Network see **Wireless Ad-hoc Network**.

ADA (Americans with Disabilities Act) 1990 federal legislation that prohibits discrimination and ensures equal opportunity for persons with disabilities in employment, government services, public accommodations, commercial facilities, and transportation. See www.ada.gov for more information.

Adaptive Multi-Rate a type of quality enhancing video codec.

ADC (Automated Data Collection) systems for collecting data at certain locations or process steps using bar codes, **RFID**, or other technologies.

Addressable a property of a device that allows it be identified and manipulated based on an identifier rather than a network location or connection.

Addressable Fire Alarm fire alarm devices that are monitored by polling individual devices or groups of devices by address. Contrast to **supervised** devices.

Adjustable Frequency Drive see **Variable Frequency Drive**.

ADPCM Adaptive Differential Pulse Code Modulation, a more efficient form of PULSE CODE MODULATION that encodes values as the difference between the current and prior sample and varies the sampling interval based on the rate of change in the signal.

ADR (Automatic Demand Response) or **Automated Demand Response,** a system that can automatically reduce electricity load (demand) in response to changes in energy availability or pricing.

ADS (Automatic Detection System) a hardware and software system that can identify events of interest within a video stream. For example, recognizing when a person is trying to jump or climb a fence.

ADSL Asymmetric Digital Subscriber Line, a form of **DSL** that offers higher speeds for DOWNLOAD than UPLOAD.

Adsorption a chemical process where a substance that was in solution in a gas or liquid accumulates on the surface of a solid or liquid. For example, silica gel is used adsorb water for drying and activated carbon is used to adsorb organic substances.

Advanced Access Content System (AACS), the content protection scheme used for video on DVD.

Advanced Application Controller an application controller **ASC** with additional alarming and scheduling capabilities.

Advanced Audio Coding standard for lossy COMPRESSION and encoding of digital audio. AAC usually achieves better sound quality than MP3 format when compared at the same bitrate, especially for bitrates below about 100 kbit/s.

Advanced Systems Format or **Advanced Streaming Format,** a Microsoft proprietary audio/video container format for streaming media as part of the Windows Media framework.

AEC Architecture, Engineering, and Construction firm.

AED (Automated External Defibrillator), a portable electronic device that diagnoses and treats potentially life threatening cardiac arrhythmias that may lead to cardiac arrest by applying electrical signals that stop the arrhythmia and allow the heart to re-establish an effective rhythm. An AED is designed for use by a layperson (ideally with CPR or FIRST RESPONDER emergency medical training) and is more limited than the defibrillators used by professionals.

Aero the graphical user interface style introduced with Microsoft Windows Vista.

AES Advanced Encryption Standard or Rijndael, is a block cipher adopted as a US government standard (FIPS 197).

AES/EBU (Audio Engineering Society/European Broadcasting Union) a group developing standards for audio technology; see www.aes.org for more information. AES/EBU is frequently used to refer to their **AES3** standard.

AES3 the **AES/EBU** standard for interconnecting digital audio equipment formalized as IEC 60958 and very similar to **S/PDIF**.

AEX Automating Equipment information exchange with XML, specifications based on **XML** to support the design, procurement, delivery, operation and maintenance of engineered equipment. Organized by FIATECH, see www.fiatech.org/projects/idim/aex.htm for more information.

AFD (Adjustable Frequency Drive) see VARIABLE FREQUENCY DRIVE.

AFDA (Automatic Fire Detection Apparatus) see AUTOMATIC FIRE DETECTION SYSTEM.

AFDS (Automatic Fire Detection System) sensors and systems to detect smoke, flames, heat, or combustion byproducts indicating a fire.

AFF Above Finished Floor.

AFUE (Annual Fuel Utilization Efficiency) the annualized average efficiency of a fuel-fired appliance, taking into account the effect of on-off operation. The higher the AFUE the lower the operating costs.

AHJ (Authority Having Jurisdiction) general purpose way of referring to the governmental organization responsible for building codes, standards and inspection.

AHU (Air Handling Unit) or **Air Handler,** a device used as part of a HVAC system that includes a blower, heating and/or cooling elements, filters, sound attenuators, and dampers. Air handlers connect to ductwork that distributes the conditioned air through part of a building, and returns it to the air handler and may include the ability to mix in OUTSIDE AIR and exhaust air from the building.

AHU Zone the portion of a building served by a single **AIR HANDLING UNIT.**

AI (Analog Input) a control device input connection capable of sensing a voltage level (**ANALOG VALUE**) within a specified range and converting it to a number or digital value. This is also known as analog to digital conversion.

AIA (American Institute of Architects) a professional organization for architects in the United States involved in advocacy, education, and standards. See www.aia.org for more information.

Air Conditioning the process of conditioning air by transferring heat and humidity from one medium to another.

Air-Conditioning and Refrigeration Institute a manufacturer's trade association; see www.ari.org for more information.

Air Handling Unit or **Air Handler,** a device used as part of a HVAC system that includes a blower, heating and/or cooling elements, filters, sound attenuators, and dampers. Air handlers connect to ductwork that distributes the conditioned air through part of a building, and returns it to the air handler and may include the ability to mix in OUTSIDE AIR and exhaust air from the building.

Air Leakage the rate of air infiltration around a window, door, or skylight in the presence of a specific pressure differential measure in units of cubic feet per minute per square foot of frame area (CFM/ft²).

AJAX Asynchronous Javascript And XML, a programming technique for Web pages where a Javascript program running in the user's browser requests data from the server on an as-needed basis in the background in order to provide a more responsive user interface.

Alarm a signal indicating a change in the state or status of a monitored system or device; an alarm may indicate an abnormal condition or some other status change. This may include burglar or intrusion alarms, fire and safety alarms, silent alarms, or equipment alarms.

Alarm and Event Management an application system designed to collect and process alarm and event messages from one or more devices or systems.

Alarm Association the process of recognizing related alarms or alarms with common causes in order to minimize the number of extraneous or NUISANCE ALARMS.

Alarm Receiving Center a system or location that receives alarm information from other locations as pat of an ALARM TRANSMISSION SYSTEM. See CENTRAL STATION.

Alarm Transmission Equipment part of an ALARM TRANSMISSION SYSTEM transporting alarm information between locations.

Alarm Transmission System a system for transporting alarm information between locations using the telephone network or other communications media.

Alert a message reporting an asynchronous event. Alert is a more general term than **ALARM** as the reported event may not be an exception or require action.

Algorithm a procedure or set of well-defined instructions for accomplishing some task.

Alien Crosstalk signal coupling between adjacent cables or between adjacent links or channels.

Alternative Fuel any non-**CONVENTIONAL FUEL** such as biodiesel, ethanol, butanol, hydrogen, methane, vegetable oil, biomass, etc.

AM (Amplitude Modulation) a signaling technique where information is transmitted as changes in the size or amplitude of a signal. AM radio uses amplitude modulation signaling in the 520–1,710 kHz range.

Ambient Art images designed to be displayed on idle televisions and video monitors; may include images of paintings, photos, computer generated effects, video loops (for example a fireplace or a fish tank), etc.

American Institute of Architects a professional organization for architects in the United States involved in advocacy, education, and standards. See www.aia.org for more information.

Americans with Disabilities Act 1990 federal legislation that prohibits discrimination and ensures equal opportunity for persons with disabilities in employment, government services, public accommodations, commercial facilities, and transportation. See www.ada.gov for more information.

AML Anti-Money Laundering.

Amplitude Modulation a signaling technique where information is transmitted as changes in the size or amplitude of a signal. AM radio uses amplitude modulation signaling in the 520–1,710 kHz range.

AMR ADAPTIVE MULTI-RATE codec or **Automated Meter Reading**.

Analog Input a control device input connection capable of sensing a voltage level (**ANALOG VALUE**) within a specified range and converting it to a number or digital value. This is also known as analog to digital conversion.

Analog Output a control system output that can be set to a specific voltage level within a specified range; also known as digital to analog conversion. Contrast to **BINARY OUTPUT** which has only two possible values.

Analog Telephone Adapter (ATA), a device that connects one or more standard analog telephones to a digital and/or non-standard telephone system such as a **VOICE OVER IP** network.

Analog Value a signal or measurement that may take on any one of a series of different values within a specified range (subject to resolution limits); for example, a temperature reading. Contrast to a **BINARY VALUE** which is either off or on.

ANI (**Automatic Number Identification**), a telephone network feature that permits subscribers to capture the telephone number of the calling party.

Annual Fuel Utilization Efficiency the annualized average efficiency of a fuel-fired appliance, taking into account the effect of on-off operation. The higher the AFUE the lower the operating costs.

Annunciator a fire and life safety system device that makes a noise or flashes lights to announce an alarm. See **NOTIFICATION APPLIANCE**.

ANSI American National Standards Institute, is a private nonprofit organization that administers American National Standards and coordinates international standards activities for the US.

ANSI/ASHRAE 135 (ISO 16484-5) the official **BACNET** standard, current version is ANSI/ASHRAE Standards 135-2004, BACnet.

ANSI/CEA 709.1 the official standard for **LonTalk,** a control protocol originally developed by Echelon.

ANSI/TIA/EIA-862 official standard for **TIA-862.**

Anticipation foreknowledge, a control strategy where action is taken in advance. For example, heating or cooling a room in based on its expected usage.

AO (Analog Output) a control system output that can be set to a specific voltage level within a specified range; also known as digital to analog conversion. Contrast to **Binary Output** which has only two possible values.

AoIP Alarm over IP, the ability to monitor **Alarm** conditions over an IP network either through alarm systems designed for use on IP networks or adapters that enable the use of an IP network in place of telephone lines or dedicated circuits for existing alarm control panels.

AP Access Point or Accounts Payable (A/P).

APC Advanced Process Control.

APDU Application Protocol Data Unit.

API (Application Programming Interface), a specification for the interface that an **Operating System** or **Application** software makes available to other application programs for purposes of sharing data, sharing functionality, allowing customization or control of the software, etc. An API can be implemented using many different technologies including functions or procedure calls, **remote procedure call, Web Services,** etc. For example, a programmer could use the Google Maps API to display a street map or satellite images of each property.

APPA an association of educational facilities professionals (formerly the Association of Physical Plant Administrators). For more information see www.appa.org.

Applet a small software application or component that runs within a larger program like a **WEB BROWSER**.

Application or **Application Program**, any software that supports a user task or business function such as word processing or accounting. Contrast to systems programs that are part of the **OPERATING SYSTEM**.

Application Programming Interface (API), a specification for the interface that an **OPERATING SYSTEM** or **APPLICATION** software makes available to other application programs for purposes of sharing data, sharing functionality, allowing customization or control of the software, etc. An API can be implemented using many different technologies including functions or procedure calls, **REMOTE PROCEDURE CALL**, **WEB SERVICES**, etc. For example, a programmer could use the Google Maps API to display a street map or satellite images of each property.

Application Server a **SERVER** designed to run application software; contrast to **MEDIA SERVER** or **DATABASE SERVER**. Web or client server applications are sometimes partitioned so that different parts of the system run on different types of servers that are tuned for different functions.

Application Service Provider company that provides access to software applications as a service over the Internet.

Application Specific Controller a controller designed to support one application at a time; sometimes used to mean a controller preconfigured for a specific application.

APS Advanced Planning & Scheduling, a manufacturing management process that optimizes materials and production capacity allocation based on demand.

AR or **A/R**, Accounts Receivable.

ARC (Alarm Receiving Center) a system or location that receives alarm information from other locations as part of an **ALARM TRANSMISSION SYSTEM**. See **CENTRAL STATION**.

Archive 1) a *place* or system for storing information; 2) the *process* of storing information, 3) *contents* of an archive, the stored information.

ARCnet Attached Resource Computer NETwork, an early local area networking technology used for building automation systems and one of the physical media supported by BACnet. See www.arcnet.com for more information.

ARI (Air-Conditioning and Refrigeration Institute) a manufacturer's trade association; see www.ari.org for more information.

ARI 550 Air Conditioning and Refrigeration Institute Standard 550-1992: Standard for Centrifugal and Rotary Screw Water-Chilling Packages or ARI Standard 550/590-1998: Standard for Water Chilling Packages Using the Vapor Compression Cycle; standard definitions, nomenclature, proper refrigerant designations, rating and testing of chillers used in comfort cooling applications.

Arm to turn on an alarm or intrusion detection system.

Artifact a flaw in a still or video image caused by lossy COMPRESSION or other sources. May appear as noise, sparkle, jagged edges, rainbow effects or video jitter.

Artificial Reasoning or Automated Reasoning, software that allows computers to reason completely or nearly completely automatically.

AS (Autonomous System) 1) a *self-healing* system that has the ability to detect and automatically compensate for problems; 2) a *robot, vehicle,* etc. that senses its environment and has the ability to make decisions and take action in pursuit of its objectives; 3) an *Internet Autonomous System* is a collection of IP networks and routers that presents a common routing policy to the Internet and is assigned an AS Number by **IANA**.

ASC (Application Specific Controller) a controller designed to support one application at a time; sometimes used to mean a controller precon-figured for a specific application.

ASCII American Standard Code for Information Interchange, a standard character encoding that supports only the English alphabet.

ASD Authorized System Distributor, typically a control system distributor that has been trained and authorized by the manufacturer.

ASF (**Advanced Systems Format**) or **Advanced Streaming Format,** a Microsoft proprietary audio/video container format for streaming media as part of the Windows Media framework.

ASHRAE the American Society of Heating, Refrigerating and Air-Conditioning Engineers, an international membership organization. For more information see www.ashrae.org.

ASIC Application Specific Integrated Circuit, an integrated circuit that is customized for a particular use.

ASIS or ASIS International, an association dedicated to increasing the effectiveness and productivity of security professionals, see www.asisonline.org for more information.

ASM Abnormal Situation Management, a research and development consortium concerned about the negative effects of industrial plant incidents. See www.asmconsortium.com for more information.

ASN Advanced Shipment Notice, an electronic notification of pending delivery or an electronic bill of lading.

ASN.1 Abstract Syntax Notation One, a flexible and efficient notation for describing and encoding data structures in computing and telecommunications networks. The current standard is part of X.680.

ASP ACTIVE SERVER PAGES, APPLICATION SERVICE PROVIDER or Average Selling Price.

Aspect Ratio the ratio of image width to height, used to characterize displays and ducts. Traditional TV or computer screens are 4:3 (1.33:1), HDTV or widescreen computers are 16:9 (1.78:1), movies may be as wide as 2.4:1.

ASRS or **AS/RS**, Automatic Storage & Retrieval Systems, applications that manage automatically depositing and retrieving loads from defined storage locations. Various systems of this type are used in manufacturing, warehouses, libraries, etc.

Asset Control see ASSET MANAGER.

Asset Manager a person or system responsible for physical or logical resources; depending on the context, an asset may be a building, equipment within a building, a movable physical asset like a piece of furniture or equipment, or a digital asset like an image or a video. Physical asset management systems designed to track portable assets may support **RFID**, bar codes, or other technologies. Managing an asset may include financial accounting, routine maintenance, inspections, certifications, condition, etc.

Asset Tracking see ASSET MANAGER.

ASTM ASTM International, originally the American Society for Testing and Materials, is an organization that develops technical standards for materials, products, systems, and services. See www.astm.org for more information.

ASTS (**Automatic Static Transfer Switch**) a device that provides both AUTOMATIC TRANSFER SWITCH and STATIC TRANSFER SWITCH functions.

ASV Approved Scanning Vendor, a firm certified to provide security scanning service by the PCI Security Standards Council. For more information see www.pcisecuritystandards.org.

Asynchronous Data Transmission data transmission that does not require synchronization between the communicating devices and instead uses start and stop bits transmitted as part of the data stream to indicate the presence of data. Asynchronous is simpler but less efficient than SYNCHRONOUS DATA TRANSMISSION because of the overhead added by the extra bits.

Asynchronous Transfer Mode (ATM), a protocol for efficient transport of constant-rate and bursty information over broadband digital networks. The ATM stream consists of fixed-length cells each containing a 5-byte header and a 48-byte payload.

ATA (Analog Telephone Adapter), a device that connects one or more standard analog telephones to a digital and/or non-standard telephone system such as a VOICE OVER IP network.

ATC AUTHORIZED TESTING CENTER or AUTOMATIC TEMPERATURE CONTROL.

ATE Automatic Test Equipment or ALARM TRANSMISSION EQUIPMENT.

ATM ASYNCHRONOUS TRANSFER MODE or Automatic Teller Machine.

Atomic Transaction a TRANSACTION that must be processed on a basis of either making all of the DATABASE updates required for the transaction or backing out all of the updates. See also ACID.

ATP Available-To-Promise, the inventory that is available to fulfill orders, specifically the on-hand inventory minus inventory allocated for back-orders and reserved inventory.

ATS ALARM TRANSMISSION SYSTEM or AUTOMATIC TRANSFER SWITCH.

ATSC Advanced Television Systems Committee, group developing DIGITAL TELEVISION standards for the US that have also been adopted by other countries. For more information see www.atsc.org.

ATV or **Advanced TeleVision**, see DTV and ATSC.

Audible Notification Appliance a NOTIFICATION APPLIANCE device that alerts by the sense of hearing.

Audio Engineering Society/European Broadcasting Union a group developing standards for audio technology; see www.aes.org for more information. AES/EBU is frequently used to refer to their **AES3** standard.

Audiovisual or **Audio-Visual**, technology for providing and controlling sound, video, lighting, display and projection systems, recording, sound masking, etc.

Authenticate the act of establishing or confirming someone or something as authentic or verifying the identity of a message sender.

Authentication, Authorization, and Accounting can be either a server providing these functions or the protocol for accessing and sharing this information as defined in RFC-2903 and related documents.

Authority Having Jurisdiction general purpose way of referring to the governmental organization responsible for building codes, standards and inspection.

Authorized Testing Center a testing facility that has been approved or authorized by a standards group to do compliance testing.

Auto Discovery a process and technology that enables systems elements to automatically detect other systems elements. For example, some BACNET controllers can auto discover new devices that are added to the network.

Automated Attendant a telephone system feature providing directory, dial-by-name, and other functions in place of a human operator.

Automated Blinds vertical or horizontal louvered devices used to control the amount of external light entering a space. See also POWER-SHADES, ROLLER SHADE.

Automated Data Collection systems for collecting data at certain locations or process steps using bar codes, **RFID,** or other technologies.

Automated External Defibrillator (AED), a portable electronic device that diagnoses and treats potentially life threatening cardiac arrhythmias that may lead to cardiac arrest by applying electrical signals that stop the arrhythmia and allow the heart to re-establish an effective rhythm. An AED is designed for use by a layperson (ideally with CPR or FIRST RESPONDER emergency medical training) and is more limited than the defibrillators used by professionals.

Automated Parking computerized parking systems ranging from garage space and availability tracking systems to robotic car storage structures.

Automatic Call Distributor a **PBX** or other telephone switching system with special features for handling large volumes of incoming calls and queuing them for multiple agent groups or other resources.

Automatic Demand Response or **Automated Demand Response,** a system that can automatically reduce electricity load (demand) in response to changes in energy availability or pricing.

Automatic Detection System a hardware and software system that can identify events of interest within a video stream. For example, recognizing when a person is trying to jump or climb a fence.

Automatic Fire Alarm System a fire detection system that will automatically detect and annunciate the presence of fire based on sensing one or more products of combustion. Annunciation is through a FIRE ALARM NOTIFICATION SYSTEM.

Automatic Fire Detection Apparatus see AUTOMATIC FIRE DETECTION SYSTEM.

Automatic Fire Detection System sensors and systems to detect smoke, flames, heat, or combustion byproducts indicating a fire.

Automatic Number Identification (ANI), a telephone network feature that permits subscribers to capture the telephone number of the calling party.

Automatic Static Transfer Switch a device that provides both AUTOMATIC TRANSFER SWITCH and STATIC TRANSFER SWITCH functions.

Automatic Temperature Control (ATC), control systems for heating, ventilating and air conditioning. An ATC Contractor furnishes and installs **HVAC** control systems and may also be responsible for COMMISSIONING.

Automatic Transfer Switch a device that transfers electrical loads between power sources automatically if one source fails.

Autonomous System 1) a *self-healing* system that has the ability to detect and automatically compensate for problems; 2) a *robot, vehicle*, etc. that senses its environment and has the ability to make decisions and take action in pursuit of its objectives; 3) an *Internet Autonomous System* is a collection of IP networks and routers that presents a common routing policy to the Internet and is assigned an AS Number by **IANA**.

Autotrack the ability of an imaging system to automatically track a moving item of interest.

AV or **A/V**, ANALOG VALUE or AUDIOVISUAL.

AV Receiver an audio/video receiver typically including an FM tuner, preamplifier, multi-channel amplifier, input and output switching for TV and speakers.

Availability the proportion of time that a system or service is available for use. See Fig. 4 on page 281 for definitions.

Avatar the incarnation or representation of a being in another realm or a person's assumed identity and shape in an online game. See for example, SECOND LIFE.

AVC Advanced Video Coding, high compression video codec standardized as H.264 or MPEG-4 Part 10.

AVI Audio Video Interleaved, a multimedia container file format introduced by Microsoft.

AVM Automated Valuation Model.

AVSP Audio Video Service Provider.

AWG American Wire Gauge, standard method of indicating wire diameter with higher numbers indicating smaller diameters.

B2B Business-to-Business.

B2C Business-to-Consumer.

B-AAC BACnet Advanced Application Controller, device profile for an
application controller with more capabilities than **B-ASC**, see **AAC**.

B-ASC BACnet Application Specific Controller, device profile for an
application controller.

B-BC BACnet Building Controller, the **BACNET** device profile for a
controller.

B-OWS BACnet Operator Workstation, the **BACNET** device profile for a
control system user interface.

B-SA BACnet Smart Actuator, device profile for an actuator.

B-SS BACnet Smart Sensor, device profile for a sensor.

Back Channel information flow from the receiver to the sender providing feedback on quality of service.

Back-of-House the parts of a facility that are behind the scenes and not open to the public or patrons; contrast to FRONT-OF-HOUSE.

Backbone Network the top level of any hierarchical communications network, typically the backbone contains the fastest links, also known as Core Network.

BACnet Building Automation and Control Networking protocol developed by ASHRAE and formalized by ANSI and ISO. Current version is ANSI/ASHRAE Standard 135-2004, BACnet or ISO 16484-5:2007.

BACnet/IP standard for BACnet networks incorporating devices that use TCP/IP protocols over ETHERNET.

BACnet/WS BACnet Web Services, addendums to the BACNET standard for WEB SERVICES support.

BACnet Broadcast Management Device (BBMD), a specialized router for BACnet broadcast messages used to forward broadcast messages between IP subnets or to distribute broadcast messages within subnets that do not allow MULTICASTING.

BACnet Device Profile Annex L of the BACNET standard defines generic device types and includes a list of **BIBB** capabilities each device might support. Profiles include operator workstation (**B-OWS**), building controller (**B-BC**), application controllers (**B-AAC** and **B-ASC**), etc.

BACnet Ethernet standard for BACnet transmission over Ethernet without using TCP/IP.

BACnet MS/TP or **BACnet MSTP**, standard for BACnet Master-Slave Token Passing protocol over a multi-drop **RS-485** serial link, sometimes called an MS/TP LAN.

BACnet PTP or **BACnet Serial,** standard for BACnet transmission over a serial **RS-232** link; sometimes used to bridge networks between buildings or locations.

BACS Building Automation and Control Systems, see **BUILDING AUTOMATION SYSTEM.**

Balanced Scorecard a framework for evaluating organizational performance based on vision and strategy. This framework was originally introduced by Kaplan and Norton in 1992 and has been adapted for a wide variety of applications.

Ballast an electrical device that provides the proper starting and operating power for a fluorescent, neon, or **HIGH-INTENSITY DISCHARGE** light. See also **HIGH POWER FACTOR BALLAST** and **NORMAL POWER FACTOR BALLAST.**

Bandwidth the volume of traffic a network connection can handle.

BAS (Building Automation System) the general class of monitoring and control systems used in buildings.

Base Building or core and shell, the overall building structure typically completed prior to tenant improvements, includes exterior walls, core (elevators, stairs, bathrooms or plumbing stubs, shafts and risers) and central mechanical and electrical systems.

Basel II or **International Convergence of Capital Measurement and Capital Standards Revised Framework,** international standards for banking operations and risk management that require tighter integration between systems and more-sophisticated tools for identifying and managing operational risks including risks to systems and facilities.

Batch Processing a computer processing job that runs without user interaction.

BBMD (BACnet Broadcast Management Device), a specialized router for BACnet broadcast messages used to forward broadcast messages between IP subnets or to distribute broadcast messages within subnets that do not allow MULTICASTING.

BC Building Controller.

BCP (Business Continuity Plan) an advance plan for operating or restoring a business or other organization following a disaster or other operational disruption.

BD BLU-RAY DISC or BUILDING DISTRIBUTOR.

BDA (Bi-Directional Amplifier), a specialized amplifier that supports sending and receiving radio signals used as part of a cellular telephone network repeater or wireless signal boosters for other types of networks.

Beam Detector or **Beam Smoke Detector**, a type of smoke detector that uses one or more light transmitters and sensors to detect the formation and thickening of a smoke layer below the ceiling. Beam detectors are typically used in atriums and other large spaces and may be installed in a vertical or horizontal grid.

BEEP BOMA Energy Efficiency Program.

BEI Biological Exposure Index, guidelines for industrial hygienists in evaluating safe levels of exposure to chemical substances and physical agents in the workplace.

Best Practice methodology that provides the best path to the desired result based on research and experience.

BHAG Big Hairy Audacious Goals.

BI BINARY INPUT or BUSINESS INTELLIGENCE.

Bi-Directional Amplifier (BDA), a specialized amplifier that supports sending and receiving radio signals used as part of a cellular telephone network repeater or wireless signal boosters for other types of networks.

BIA Business Impact Analysis, the part of a **Business Continuity Plan** that evaluates the risks and financial impact of system disruption due to a disaster against different cost-to-recover solutions and recommends appropriate risk mitigation measures.

BIBB BACnet Interoperability Building Block, a set of common interoperability functions defined in Annex K of the standard. Functions are grouped by area: data sharing, alarms and events, scheduling, trending, and device management.

BICSI (Building Industry Consulting Services International), an association of people involved in the design and installation of network infrastructure or **ITS** distribution systems. For more information see www.bicsi.org.

Big Box Retailer a retail chain with a number of large stores; 50,000 square feet is the lower cut-off with typical store sizes from 90,000 to 200,000 square feet.

BIM (Building Information Model) a computer representation of the physical and functional characteristics of a facility. Models are created in **CAD** tools and may interoperate with other software for engineering analysis or other processing.

Binary Input an input that can only have one of two possible values; on or one and off or zero; contrast to **Analog Input**.

Binary Output an output from a system or device with two possible values corresponding to on or one and off or zero.

Binary Value a signal or measurement that may have only one of two possible values; these values may be described as 0 or 1 or off and on, depending on the context. Contrast to **Analog Value**.

Biological Agent any type of virus, spore or bacteria released with the intent to cause harm.

Biometric physical property of a person used to verify identification. Examples include finger prints, iris scan, voice properties, hand geometry, etc.

Bioreader a BIOMETRIC reader device for a fingerprint or other characteristic.

BIQ (Building Intelligence Quotient) an online tool for rating buildings. See www.buildingintelligencequotient.com for more information.

Bitmap a simple type of image containing pixels with compression or encoding.

Bitrate the number of bits used to represent a second of audio or video.

Black Box Testing a testing approach that assumes no prior knowledge of the system or infrastructure to be tested.

Black Level light level in the darkest part of a video image.

Black Water or **Blackwater,** water containing fecal matter and urine. Contrast to GRAY WATER.

Blacklist or **Black List,** a type of ACCESS CONTROL LIST naming users or systems that are denied access; contrast to WHITE LIST.

Blade Server a high density server in a small form-factor that plugs into an enclosure and connects to a shared backplane that provides power and communications.

Blinds see AUTOMATED BLINDS, POWER-SHADES, SUN BLINDS.

BLIS (Building Lifecycle Interoperable Software) a coordination project to support the implementation of **IFC** specification in software products. See www.blis-project.org for more information.

Blog short for Web log, a web-based publication consisting mostly of periodic articles shown in reverse chronological order.

Blu-ray Disc a standard for a higher density optical disk currently capable of storing 50 **Gigabytes**, enough for nine hours of HD video or 23 hours of standard-definition video.

Bluejacking slang term for sending anonymous text messages to other phone users via a **Bluetooth** wireless link.

Bluetooth standard for a wireless personal area network, also known as IEEE 802.15.1. Connects devices like personal digital assistants, mobile phones, laptops, PCs, printers and digital cameras via short range radio frequency links.

BMA the BACnet Manufacturer's Association, organization for **BACnet** promotion and education now known as BACnet International. See www.bacnetassociation.org for more information.

BMCS Building Management Control System.

BMS Building Management System, a software application that can monitor and control the automated systems in a building; sometimes used as an alternate term for **Building Automation System**.

BNC a type of bayonet mount connector used for coaxial cable.

BO (Binary Output) an output from a system or device with two possible values corresponding to on or one and off or zero.

BOC Bell Operating Company or **Building Operations Center**.

BOL Bill of Lading.

Bollard a vertical post designed to protect vulnerable areas of a facility from vehicles. Bollards may be fixed, moveable, or automatic.

BOM Bill of Materials.

BOMA Building Owners and Managers Association, an organization of commercial real estate professionals, see www.boma.org for more information.

BON Building Optical Network, a shared high speed backbone or CORE network that supports both BUILDING AUTOMATION SYSTEMs and tenant communications; see also STRUCTURED CABLING SYSTEM.

BOOM Build, Own, Operate, Maintain.

Boot **Booting** or **Booting Up**, the bootstrapping process that takes place when a computer system is turned on or reset. Typically this includes a POWER ON SELF TEST, loading the first part of the OPERATING SYSTEM software called the Boot Loader (from the Boot Record or Boot Sector on disk) and starting the Boot Loader which then loads and starts the rest of the operating system.

BOOT Build, Own, Operate, and Transfer, the Canadian term for BUILD-OPERATE-TRANSFER financing.

BOT (**Build-Operate-Transfer**) a type of project financing where a private entity receives a franchise from a public sector agency to finance, design, construct, and operate a facility for a period of time, after which the franchise is transferred back to the public sector agency.

Bottleneck the element of any system with the highest utilization that restricts throughput.

BPA Business Process Automation.

BPE Business Process Engineering.

BPM (**Business Process Management**) the methods, techniques and systems used to design, implement, control, and analyze operational business processes involving people, organizations, applications, documents and other sources of information. See also WORKFLOW.

BPO Business Process Outsourcing.

BPR Business Process Reengineering.

BRFL Building Fire Research Laboratory, part of **NIST**, see www.bfrl.nist.gov for more information.

BRI Basic Rate Interface, or **BUILDING RELATED ILLNESS**.

Brightness light level of an overall image. On an electronic display the brightness control sets the **BLACK LEVEL** and the **CONTRAST** control adjusts the brightness of the white level.

Broadband high speed data communications, generally defined as faster than 56 kilobits per second.

Broadcast to transmit broadly, usually simultaneously to all recipients.

BSOD Blue Screen of Death, a Microsoft Windows error display with white text on a blue background that indicates a serious system error.

BSP Best Security Practices or Business Services Provider.

BTL **BACNET** Testing Laboratory, see www.bacnetinternational.org for more information.

BTU British Thermal Unit, a unit of energy equal to 1060 joules or the amount of heat required to raise the temperature of 1 pound of water by 1 degree Fahrenheit.

BTUH BTU per Hour, the basic small unit for measuring the rate of heat transfer.

Buffer an amount of storage or space designed to provide some operational flexibility or cushion. Computer buffers are limited, leading to the possibility of buffer overflow attacks.

Build-Operate-Transfer a type of project financing where a private entity receives a franchise from a public sector agency to finance, design, construct, and operate a facility for a period of time, after which the franchise is transferred back to the public sector agency.

Building Automation System the general class of monitoring and control systems used in buildings.

Building Code a set of rules designed to protect public health, safety and general welfare by specifying a minimum acceptable level of safety for buildings and other structures. Building codes may also impose requirements for energy efficiency, INDOOR AIR QUALITY, etc. There are many different building codes published by different organizations; a building code becomes law for a jurisdiction when formally enacted by the AUTHORITY HAVING JURISDICTION.

Building Distributor IEC terminology for a INTERMEDIATE CROSS-CONNECT.

Building Envelope the walls and roof enclosing a building.

Building Industry Consulting Services International (BICSI), an association of people involved in the design and installation of network infrastructure or ITS distribution systems. For more information see www.bicsi.org.

Building Information Model a computer representation of the physical and functional characteristics of a facility. Models are created in CAD tools and may interoperate with other software for engineering analysis or other processing.

Building Intelligence Quotient an online tool for rating buildings. See www.buildingintelligencequotient.com for more information.

Building Lifecycle Interoperable Software a coordination project to support the implementation of IFC specification in software products. See www.blis-project.org for more information.

Building Operations Center the organization and computer applications supporting the maintenance and operation of buildings.

Building Operator a person with knowledge of and control over HVAC and other building systems.

Building Related Illness the term used when symptoms of diagnosable illness are identified and can be attributed directly to airborne building contaminants. See also SICK BUILDING SYNDROME.

buildingSMART an umbrella term for IAI efforts at building industry process improvements. See www.iai-na.org/bsmart/ for more information.

Built Environment a concept that includes all constructed entities and spaces created or modified by human intervention.

Burn In operating and monitoring new equipment for an initial period during which failures are more likely.

Bus 1) *electrical bus* or **busbar** is a heavy duty conductor used to distribute electric power to components of a system; 2) a *computer bus* is used to connect together components of a system, for example a PCI bus; 3) a *communications bus* is a linear network consisting of a common communications path shared by a number of nodes, for example USB; 4) a *software bus* is a standardized interface designed to facilitate the process of integrating multiple system components, for example ENTERPRISE SERVICE BUS; 5) a *vehicle* for multiple passengers.

Business Continuity Plan an advance plan for operating or restoring a business or other organization following a disaster or other operational disruption.

Business Intelligence processes and capabilities for turning data (frequently from a DASHBOARD or DATA WAREHOUSE) into actionable information that helps users make better business decisions.

Business Process Management the methods, techniques and systems used to design, implement, control, and analyze operational business processes involving people, organizations, applications, documents and other sources of information. See also **WORKFLOW**.

BV (Binary Value) a signal or measurement that may have only one of two possible values; these values may be described as 0 or 1 or off and on, depending on the context. Contrast to **ANALOG VALUE**.

Byte or **Octet,** a unit of information storage containing eight binary digits or bits. One byte generally represents one character in a Western language; Eastern languages and **UNICODE** may require multiple bytes per character.

Bytecode a binary representation of an executable program generated by some types of compilers, similar to object code.

C

C&I Commercial and Industrial.

C++ an **OBJECT ORIENTED** programming language based upon the C **LANGUAGE**.

C/I MLS (Commercial/Industrial MLS) a **MULTIPLE LISTING SERVICE** for commercial and industrial properties. Listing a property typically implies an offer of compensation to other participants.

C40 a group of the world's largest cities committed to working together on climate change programs. For more information see www.c40cities.org.

C-Bus low voltage wired (**CAT 5**) and wireless control system for lighting and other applications.

C Language a general-purpose, procedural, computer programming language.

CA CERTIFICATE AUTHORITY, CONDITIONAL ACCESS, or Construction Administration.

CABA (Continental Automated Buildings Association), a nonprofit industry organization that promotes the development and understanding of home and building automation, see www.caba.org for more information.

Cable Fill the ratio of the cable installed in a conduit or other area compared to the theoretical maximum capacity.

Cable Ladder a type of ladder-like cable rack used to support and protect wiring or fiber optic cables in both vertical and horizontal applications.

Cable Mining the process of identifying and removing unused cables within a building. Since 2002 the NATIONAL ELECTRICAL CODE has required the removal of abandoned communications cables due to health and life safety concerns centered on **PVC** cable jackets.

Cable TV technology for providing television via radio frequency signals transmitted using optical fibers or coaxial cables as opposed to over-the-air radio broadcasting. Cable TV networks may also provide FM radio programming, high-speed Internet access, telephony and other non-television services.

Cache a local copy of data stored in a high speed memory to improve performance.

CAD CHEMICAL AGENT DETECTION, COMPUTER AIDED DESIGN software, or CONVENTIONAL AIR DISTRIBUTION.

CAE (Computer Aided Engineering) software for engineering calculations.

CAFM see COMPUTER AIDED FACILITIES MANAGEMENT.

CAGR Compound Annual Growth Rate.

Call Center a specialized facility designed to support high volume telephone activity including inbound calls, outbound calling, or both. May be equipped with special telephone systems, see **AUTOMATIC CALL DISTRIBUTOR** and **PREDICTIVE DIALER** and staffed with some number of customer service representatives.

CALS (Continuous Acquisition and Lifecycle Support) formerly Computer-aided Acquisition and Lifecycle Support, an initiative to develop standards for electronic data interchange, electronic technical documentation, and guidelines for process improvement that was started by the US Department of Defense.

CAM Common Area Maintenance, **COMPUTER AIDED MANUFACTURING** or **COMPUTERIZED ASSET MANAGEMENT**.

Campus Distributor IEC terminology for a **MAIN CROSS-CONNECT**.

CAN Controller Area Network, a serial bus interface standardized as ISO 11898-1.

CAP (Common Alerting Protocol), an open standard format for all types of alert and notification messages that is based on **XML** and supports **SAME**; CAP allows a warning message to be consistently disseminated over many warning systems to many applications, increasing warning effectiveness and simplifying the task of activating a warning for responsible officials. The CAP standard was developed by **OASIS** and has been submitted to the **ITU** for adoption.

Cap Rate or **Capitalization Rate,** the ratio between the annual cash flow produced by an asset and its capital cost; for a building this is net operating income divided by price. Cap rates measure how quickly an investment pays for itself in terms of the cash flows it generates and are also used in determining property valuation.

Capacity a general term that can mean processing power, transmission speed, or storage ability, depending on the context. See Fig. 1 on page 279 for information storage capacity measures.

Capacity Planning a business process that includes measuring and monitoring current resource utilization, estimating future resource requirements, and planning upgrades or changes required to meet future demands on a computer system, communication network, or other resource.

Capex or **Capital Expense**, the cost to acquire or upgrade assets with a useful life of more than one accounting period such as equipment, property, buildings, etc. Capital Expenses may be subject to depreciation and amortization; contrast to OPEX.

CAPI or **CryptoAPI**, CRYPTOGRAPHIC APPLICATION PROGRAMMING INTERFACE.

CAR Certification Authorization Report.

Carbon Dioxide (CO_2), a gas (at room temperature) produced by combustion, respiration, or outgassing from certain materials; outside air concentration is typically low (0.04%). Carbon dioxide is a surrogate for indoor pollutants that may cause occupants to grow drowsy, get headaches, or function at lower activity levels; ASHRAE recommends that indoor CO_2 levels not exceed 0.1% or 1000 ppm.

Carbon Monoxide (CO), an extremely toxic gas that is colorless, odorless, and tasteless and is produced by the incomplete combustion of carbon compounds. Symptoms of mild poisoning include headaches, dizziness, and flu-like effects.

Card Access an ACCESS CONTROL SYSTEM based on the use of some type of card as the identification or authentication credential.

Carpet and Rug Institute national trade association representing the carpet and rug industry. For more information see www.carpet-rug.org.

Carriage Return (CR), a control character that commands a printer or other display to position of the cursor at the first position on the current line.

Carrier Ethernet or **Carrier Grade Ethernet,** enhanced Ethernet with additional features based on the needs of telecommunications carriers; enhancements include scalability, protection (reliability and disaster recovery), QUALITY OF SERVICE, TIME DIVISION MULTIPLEXING support, and services management, etc.

CASBAA Cable And Satellite Broadcast Association of Asia, see www.casbaa.com.

Cascading Style Sheets a language used to describe and control the presentation of documents written in a markup language such as HTML or XHTML; this separates presentation formatting from content and makes it possible to have specialized presentation formats for different device or media types (screen, handheld, print), usage, or user-type. See www.w3.org/Style/CSS/ for more information.

Case Management a HELP DESK or TROUBLE TICKET support application may combine related problem reports into a single case.

Cat-1 unshielded twisted pair cabling for telephone communications.

Cat-5 a type of unshielded twisted pair Ethernet cable. Technically Cat-5 has been superseded by Cat-5e in the TIA/EIA-568-B standard. See Fig. 12 on page 288.

Cat-6 a cable standard for Gigabit Ethernet ANSI/TIA/EIA-568-B.2-1.

CATV Community Antenna Television or CABLE TV.

CB Chemical or Biological.

CBD Central Business District.

CBECS (Commercial Buildings Energy Consumption Survey) a national sample survey conducted by the US Department of Energy every four years. See www.eia.doe.gov/emeu/cbecs/ for more information.

CBO (Centralized Building Operations) the ability to monitor and operate multiple buildings from a centralized location.

CBR CHEMICAL/BIOLOGICAL/RADIATION or **CONSTANT BIT RATE.**

CBT Computer-based Training.

CC (Common Criteria), the ISO/IEC 15408 standard for computer security developed from North American and European defense and intelligence standards. Common Criteria describes a framework where users can define security requirements, vendors can implement and/or make claims about the security of their products, and testing laboratories can evaluate the products.

CCD Charge Coupled Device, a type of sensor used in digital cameras and other applications to measure brightness and color.

CCDP (Closed-Circuit Digital Photography) use of automatic digital cameras that take high resolution still images periodically or when motion is detected.

CCHP Combined Cooling, Heating, and Power, see **COMBINED HEATING AND POWER.**

CCI (Clinton Climate Initiative), a project of the Clinton Foundation intended to make a practical and measurable difference in the fight against climate change. For more information see www.ClintonFoundation.org.

CCTV (Closed-Circuit Television) video transmitted over a private network.

CD CAMPUS DISTRIBUTOR, Ceiling Diffuser, **COMPACT DISC**, or Construction Documents.

CDDI (Copper Distributed Data Interface) a version of the **FIBER DISTRIBUTED DATA INTERFACE** that uses copper twisted pair cables instead of fiber optics.

CDE (Common Desktop Environment) an integrated graphical user interface for open systems desktop computing, see www.opengroup.org/cde/ for details.

CDMA or **IS-95**, Code Division Multiple Access, a 2G digital cellular telephony standard based on a multiple access scheme where channels are divided using codes, pioneered by Qualcomm.

CDMA2000 or **3GPP2**, a hybrid 2.5G/3G cellular telephony standard based on **CDMA**.

CDSS Commercial Data Storage Standards.

CEA (Consumer Electronics Association), a consumer electronics industry association that includes manufacturers, dealers, and installers of audio, home networking/theater, mobile electronics and wireless technology. See www.ce.org for more information.

CEC (Consumer Electronic Control) a channel for remote control functions specified as part of the **HDMI** interface.

CEMS Continuous Emissions Monitoring System, a system that can determine the gas or particulate matter concentration or emission rate using pollutant analyzer measurements and convert these results into the units of the applicable emission limitation or standard.

CENELEC Committée de European de Normalization Electrotechnique or European Committee for Electrotechnical Standardization, the European standards organization for electrical engineering.

Center Channel the center or voice channel in a home theater setup.

Center for Integrated Facility Engineering Stanford University research center for virtual design and construction of Architecture Engineering Construction (AEC) industry projects. For more information seee www.stanford.edu/group/CIFE/.

Center for Resource Solutions a nonprofit organization working to build a robust renewable energy market by increasing demand and supply of renewable resources. For more information see www.resource-solutions.org.

Central Office in traditional telephone networks a central office is the building that holds the telephone switching equipment, also known as an exchange or wire center.

Central Processing Unit the part of a computer or microprocessor that does the actual calculating and processing.

Central Station the main monitoring point of a security system. See also UL CENTRAL STATION.

Central Station Alarm Association a trade association for providers, users, bureaus, and other agencies of central station protection services. See www.csaaul.org for more information.

Central Station Receiver a device that accepts connections from PREMISES EQUIPMENT for the purpose of transmitting event or alarm information to the CENTRAL STATION.

Centralized Building Operations the ability to monitor and operate multiple buildings from a centralized location.

Centrifugal Chiller a centrifugal compressor uses a rotating disk or impeller to force gas to the rim of the impeller, increasing the velocity of the gas; a diffuser converts the velocity energy to pressure energy, the chiller then uses the vapor compression cycle to chill water (or other fluids) and transfers the heat collected from the chilled water plus the heat from operating compressor to a second water loop that can be cooled by a COOLING TOWER.

CEP (Complex Event Processing) an application for processing, correlating, and managing alarm or event messages.

CER • Certified Technology Specialist

CER (Common Equipment Room), TIA-862 terminology for a shared MECHANICAL ROOM where BUILDING AUTOMATION SYSTEM controllers are located.

Certificate an official document affirming some fact. A SECURITY CERTIFICATE uses a digital signature (typically from a CERTIFICATE AUTHORITY) to bind a public ENCRYPTION KEY to an identity; for example to validate the ownership of a website. An insurance certificate shows the existence and validity of insurance coverage. Other types of certificates relate to standards compliance, credentials, etc.

Certificate Authority or **Certification Authority** (CA), 1) for a SECURITY CERTIFICATE the CA is an entity that issues DIGITAL CERTIFICATEs or encryption keys for use by other parties as part of a PUBLIC KEY INFRASTRUCTURE system. The CA acts as a trusted third party that verifies applicant credentials and attests that the key contained in the certificate belongs to the person, organization, server or other entity noted in the certificate; there may be a hierarchy of certificate authorities. 2) For a **certificate of insurance** the CA attests to the validity of the certificate. 3) *Other* uses include certification of standards compliance or validating types of certificates.

Certificate Management 1) SECURITY CERTIFICATE related business processes and systems for certificate issuance, and renewal, user management, certificate revocation, data and key recovery, see also **X.509** protocols for certificate management and **KEY MANAGEMENT SYSTEM**; 2) **certificate of insurance** management, business processes and systems to assure that all suppliers, contractors, tenants, etc. have provided appropriate and current proof of insurance coverage.

Certificate of Occupancy a document issued by a local government agency or building department certifying that a building or a portion of a building has been satisfactorily inspected, is in a condition suitable for occupancy, and how the space may be occupied.

Certified Technology Specialist (CTS), a certification offered by the ICIA (see **INFOCOMM**); CTS specialties include AUDIOVISUAL design and installation.

CFAD Conventional Flow Air Distribution, see CONVENTIONAL AIR DISTRIBUTION, contrast to **UFAD**.

CFAR Collaborative Forecasting & Replenishment.

CFC (Chlorofluorocarbons) or **Freon**, chemical compounds consisting of chlorine, fluorine, and carbon. These chemicals were used as REFRIGERANTs, propellants, and cleaning solvents before they were banned because of their effect on the ozone layer.

CFCI Contractor Furnished and Contractor Installed.

CFD Computational Fluid Dynamics, mathematical techniques for modeling fluid flow.

CFE Customer Furnished Equipment.

cfiXML Capital Facilities Industry XML, an industry organization developing open source software, testing tools, and reusable XML schemas for managing the life cycle of capital equipment. See www.cfixml.org for more information.

CFL (Compact Fluorescent Lamp) a fluorescent lamp that screws into a standard light bulb socket or plugs into a small lighting fixture. Also known as a compact fluorescent light bulb or an energy saving light bulb.

CFM Chief Facilities Manager, Continuous Flow Manufacturing, or Cubic Feet per Minute.

CFMA (Construction Financial Management Association) a professional organization for construction financial professionals and service providers. See www.cfma.org for more information.

CFR Code of Federal Regulations.

CGI (**Common Gateway Interface**), a standard protocol for interfacing application software with a WEB SERVER. CGI allows the server to pass requests from a web browser client to the application and return output from the application to the web browser.

Change Management the process of identifying, analyzing, planning, approving, coordinating, testing, and documenting changes to the CONFIGURATION of a system. This may include identifying potential problems during the change process and creating contingency plans for dealing with these problems if they arise.

Change of State (COS), a type of event support available in BACNET and other systems where instead of continuously polling devices for their state, an application can subscribe to receive a change of state event notification when the state changes. This reduces network traffic and communications processing load.

Change of Value (COV), a type of event support available in BACNET and other systems where instead of continuously polling devices for their current value, an application can subscribe to receive a change of value event notification when the value changes. This reduces network traffic and communications processing load.

Change Order a request to make a change in the schedule, process, or deliverables of a project or to change a service.

Channel or **Communications Channel**, one path for voice or signal transmission within a shared communications media. Channels may be created using time division, frequency division, coding or addressing schemes, token passing, etc.

Charge to fill to capacity. The normal amount of refrigerant in an HVAC system is called a charge, the process of filling or refilling is charging.

Charrette or **Design Charrette**, a workshop or an intense period of design activity intended to develop several possible solutions.

Chartered Institution of Building Services Engineers an international membership organization for building services professionals. For more information see www.cibse.org.

Checkvalve a mechanical device that allows fluid to flow in one direction only.

Chemical/Biological/Radiation types of terrorist threats or attacks via airborne contaminants.

Chemical Agent a poisonous substance released with the intent to cause harm.

Chemical Agent Detection technology for detecting chemical warfare agents and other chemical hazards.

Chief Security Officer high level executive responsible for physical and network security.

Chilled Beams radiant cooling and heating solutions based on moving temperature controlled water, rather than air, that can be more energy efficient and require less space compared to conventional HVAC systems. Chilled beams may also incorporate acoustic baffling, lighting and other services.

Chilled Ceiling Beam see CHILLED BEAMS.

Chiller a machine that removes heat from a liquid via a vapor-compression or absorption refrigeration cycle. The liquid may be plain water, water mixed with antifreeze, or other REFRIGERANT fluids. See also ABSORPTION CHILLER and CENTRIFUGAL CHILLER.

CHIPS see COLLABORATIVE FOR HIGH PERFORMANCE SCHOOLS.

Chlorofluorocarbons or **Freon**, chemical compounds consisting of chlorine, fluorine, and carbon. These chemicals were used as REFRIGERANTs, propellants, and cleaning solvents before they were banned because of their effect on the ozone layer.

CHP (Combined Heating and Power) or **Combined Heat and Power,** systems designed to meet the thermal and electrical needs of a facility from the same fuel source(s).

CHPS (Collaborative for High Performance Schools) group developing standard design and construction criteria for greener, healthier schools. See www.chps.net for more information.

Churn turnover or change in customers or occupants.

CHW Chilled Water.

CI Counter-Intelligence, activities to prevent an enemy from obtaining secret information.

CIBSE (Chartered Institution of Building Services Engineers) an international membership organization for building services professionals. For more information see www.cibse.org.

CIE Commercial Information Exchange, Control Indicating Equipment, or International Commission on Illumination.

CIFE (Center for Integrated Facility Engineering) Stanford University research center for virtual design and construction of Architecture Engineering Construction (AEC) industry projects. For more information seee www.stanford.edu/group/CIFE/.

CII (Construction Industry Institute) a consortium of owners, engineering and construction contractors, and suppliers working to improve the cost effectiveness of the capital facility project life cycle. See www.construction-institute.org for more information.

CIM see Common Information Model computing, Common Information Model power or Computer Integrated Manufacturing.

CIO Chief Information Officer.

CIP (Common Industrial Protocol) or **DeviceNet**, messages and services for manufacturing automation applications. See www.odva.org for more information.

Circuit Switching a type of connection where a connection between the caller and the recipient is reserved for the entire duration of the call. Contrast to PACKET SWITCHING networks.

CISC or **Complex Instruction Set Computing**, a type of processor design based on sophisticated instructions, contrast to **RISC**.

CLA (Communications, Life-safety, and Automation) the cabling infrastructure and systems for voice, data and wireless communications, audio, video, paging, intercom, DIGITAL SIGNAGE, fire detection and alarm, CCTV, ACCESS CONTROL, INTRUSION DETECTION SYSTEM, and building automation.

CLADI (Communications Life safety Automation Design Institute), a group established by **BICSI** to help members develop construction specifications for telecommunications infrastructure. See www.cladi.org for more information.

Class A office space, buildings competing for premier office users, charging above average rents, and offering high quality finishes and amenities.

Class B office space, buildings competing for a wide range of tenants with rents in the average range for the area.

Class C or Flex Class office space, buildings offering functional space at rents that are below average for the area.

Class of Service (CoS), 1) *networking*, a mechanism for tagging network traffic with a priority value in order to differentiate and prioritize data streams; 2) *voice telephony*, a mechanisms used to select phone features on a **PBX**, define line types, and give selective priority.

Clean Water Act the primary US law governing water pollution, US Code section 33 subsection 1251.

Clg Cooling.

ClgSP Cooling SETPOINT.

Client a network device that uses information provided by a SERVER. Personal computers, personal digital assistants, and other devices that provide a user interface act as clients.

Client-Server a type of DISTRIBUTED PROCESSING computer applications architecture where some processing is done in each user's device or CLIENT and databases and other information are stored in a central system known as the SERVER. At one time client-server meant PC's and a single server connected via a LAN within an office, now the term is used for all types of distributed applications including Internet services.

Clinton Climate Initiative (CCI), a project of the Clinton Foundation intended to make a practical and measurable difference in the fight against climate change. For more information see www.ClintonFoundation.org.

Clip a digital media file.

CLN Collaborative Logistics Network.

CLO Clothing Insulation, insulating effect of occupant clothing.

Clock Control System a system for setting and controlling clocks or other time display mechanisms.

Closed-Circuit Digital Photography use of automatic digital cameras that take high resolution still images periodically or when motion is detected.

Closed-Circuit Television video transmitted over a private network.

CMDB (**Configuration Management Database**) another name for a CONFIGURATION MANAGEMENT application that emphasizes the importance of the underlying database.

CMMS (Computerized Maintenance Management System) a computer application that supports and organizes maintenance and repair activities for buildings, equipment, vehicles, etc.

CMOS Complementary Metal Oxide Semiconductor, a class of integrated circuits known for lower power requirements.

CMR or **Cellular Mobile Radio,** another name for cellular telephone service.

CMS Cable Management System (see **TIA-606**), **Change Management** System, or **Configuration Management** System.

CMYK Cyan Magenta Yellow and Key (or black), a four color subtractive color model used for commercial color printing with four inks. Contrast to **RGB**, an additive color model used by most electronic devices.

CNC (Computer Numerical Control) a general term for automated machine tools.

CO Carbon Monoxide, Central Office in telephony, or **Change Order** in project management.

CO2 or **CO$_2$, Carbon Dioxide.**

COBIT Control Objectives for Information and related Technology, a framework of best practices for information technology management created by the **Information Systems Audit and Control Association** and the **IT Governance Institute,** standardized as ISO/IEC 17799:2005, and encouraged by **Sarbanes-Oxley.**

Code 1) as in **Building Code,** a standard that is an extensive compilation of provisions covering broad subject matter or that is suitable for adoption into law; 2) *software* programs as written (source code) or generated by a compiler (object code); 3) an *address* or identifier.

CODEC Compression Decompression hardware or software used to encode and decode digital audio or video signals.

Coefficient of Performance a measure of chiller energy efficiency.

Cogeneration any process that uses one energy source to generate both electricity and usable heat simultaneously. See **COMBINED HEATING AND POWER**.

COL (Common Object Library) a software framework being developed as part of ace**XML**.

Cold Chain a temperature-controlled and monitored supply chain for food, pharmaceuticals or chemicals that maintains a specified temperature range during storage and distribution activities.

Collaborative for High Performance Schools group developing standard design and construction criteria for greener, healthier schools. See www.chps.net for more information.

Collaborative Production Management a systems approach to manufacturing decision-making that incorporates real-time input from suppliers and customers.

Color Rendering Index (CRI), a **CIE** quantitative measure of the ability of a light source to reproduce colors on a scale of zero (poorest) to one hundred. A monochromatic low-pressure sodium vapor lamp is nearly zero and an incandescent light bulb is nearly one hundred.

COM COMPONENT OBJECT MODEL or Customer's Own Material.

Combined Heating and Power or **Combined Heat and Power,** systems designed to meet the thermal and electrical needs of a facility from the same fuel source(s).

Combined Heating Power and Cooling systems designed to meet the heating, cooling and electrical power generation needs of a facility from the same fuel source(s).

Comfort Index various measures designed to reflect the level of perceived satisfaction with environmental conditions in a single number. Depending on the definition and intended use this index may include combinations of heat and humidity, temperature and wind speed, or the ratio of actual and expected values.

Comma Separated Values (CSV), a file format for representing tabular data as a text file with one line per record and a comma to separate values. Some versions of this format allow quotes around fields that contain reserved characters (such as comma or line endings) or a heading row containing column labels.

Command and Control Center centralized operations center for a system. See also NETWORK OPERATIONS CENTER, OPERATIONS CONTROL CENTER.

Commercial/Industrial MLS a MULTIPLE LISTING SERVICE for commercial and industrial properties. Listing a property typically implies an offer of compensation to other participants.

Commercial Buildings Energy Consumption Survey a national sample survey conducted by the US Department of Energy every four years. See www.eia.doe.gov/emeu/cbecs/ for more information.

Commercial Information Exchange an Internet-based service that allows its users to submit, search, and display commercial real estate transaction information including listings for a specific market. Unlike a MULTIPLE LISTING SERVICE, no offer of compensation is implied by listing a property on a CIE.

Commissioning the process of preparing systems for use including configuring the systems, start up, performance testing and verification, operator training, and establishing operations and maintenance criteria. The objective of the commissioning process is to ensure that the system functions as designed.

Common Alerting Protocol (CAP), an open standard format for all types of alert and notification messages that is based on **XML** and supports **SAME**; CAP allows a warning message to be consistently disseminated over many warning systems to many applications, increasing warning effectiveness and simplifying the task of activating a warning for responsible officials. The CAP standard was developed by **OASIS** and has been submitted to the **ITU** for adoption.

Common Criteria (CC), the ISO/IEC 15408 standard for computer security developed from North American and European defense and intelligence standards. Common Criteria describes a framework where users can define security requirements, vendors can implement and/or make claims about the security of their products, and testing laboratories can evaluate the products.

Common Desktop Environment an integrated graphical user interface for open systems desktop computing, see www.opengroup.org/cde/ for details.

Common Equipment Room (CER), **TIA-862** terminology for a shared MECHANICAL ROOM where BUILDING AUTOMATION SYSTEM controllers are located.

Common Front End a system that provides access to multiple other systems from a single window or workstation.

Common Gateway Interface (CGI), a standard protocol for interfacing application software with a WEB SERVER. CGI allows the server to pass requests from a web browser client to the application and return output from the application to the web browser.

Common GUI a system providing a consistent Graphical User Interface (**GUI**) to multiple systems or applications.

Common Industrial Protocol or **DeviceNet**, messages and services for manufacturing automation applications. See www.odva.org for more information.

Common Information Model computing a standard data model for representing managed elements of an IT environment developed by DMTF as part of WBEM. See www.dmtf.org/standards/cim/ for more information.

Common Information Model power an IEC standard for the electrical power industry that allows transmission network operators to exchange information about the configuration and status of their networks. See www.cimuser.org for more information.

Common Object Library a software framework being developed as part of aceXML.

Common Return a HVAC duct, plenum, or mixing chamber that draws air from several parts of a building, allows the air to mix, and delivers it to one or more AIR HANDLING UNIT.

Common Telecommunications Room (CTR), a shared TELECOMMUNICATIONS ROOM.

Common User Interface system or software that provides a consistent user interface to multiple systems or applications from a single window or workstation.

Communications, Life-safety, and Automation the cabling infrastructure and systems for voice, data and wireless communications, audio, video, paging, intercom, DIGITAL SIGNAGE, fire detection and alarm, CCTV, ACCESS CONTROL, INTRUSION DETECTION SYSTEM, and building automation.

Communications Life Safety Automation Design Institute (CLADI), a group established by BICSI to help members develop construction specifications for telecommunications infrastructure. See www.cladi.org for more information.

Compact Disc an optical disk used to store digital data, audio, or video. A 120 mm disk can hold approximately 80 minutes of audio (700 MEGA-BYTES), an 80 mm disk holds approximately 20 minutes of audio.

Compact Fluorescent Lamp a fluorescent lamp that screws into a standard light bulb socket or plugs into a small lighting fixture. Also known as a compact fluorescent light bulb or an energy saving light bulb.

Complex Event Processing an application for processing, correlating, and managing alarm or event messages.

Compliance business processes for adhering to and documenting adherence to laws, regulations or policies; for example internal controls, audit trails, external audits, etc. Building code compliance is determined by the AUTHORITY HAVING JURISDICTION.

Component a reusable software object that contains data and supports certain services or methods that are exposed to other programs. See also OBJECT ORIENTED.

Component Object Model (COM), a Microsoft framework for software components that enables dynamic object creation and communications between objects.

Component Video video signal broken down into one luminance channel and two color channels. Theoretically component video offers more detail than S-VIDEO or COMPOSITE VIDEO.

Composite Video a video signal with both color (chrominance) and intensity (luminance) on the same cable.

Compression processing data to remove redundant information and reduce the number of bits required for storage or transmission. Some compression methods are 'loss-less' reversible (see for example ZIP compression) and others are 'lossy' and degrade the data (see for example JPEG compression which reduces image quality).

Computer Aided Design software tools for architectural and technical design.

Computer Aided Engineering software for engineering calculations.

Computer Aided Facilities Management applications to organize and
manage facilities information including building and property information, floor plans and space usage, employee and occupancy data, telecom
infrastructure, physical assets, business continuity and safety information,
etc.

Computer Aided Manufacturing general term for the use of computer
applications to improve manufacturing, includes product lifecycle
management, **CAD** and **CNC** tools.

Computer Integrated Manufacturing manufacturing supported
by computer applications including COMPUTER AIDED DESIGN,
COMPUTER AIDED MANUFACTURING and other business applications
and databases.

Computer Numerical Control a general term for automated machine
tools.

Computer Room an area equipped with power, cooling and cable management facilities to support computer equipment. May also have raised
floor, specialized fire suppression systems, etc.

Computer Room Air Conditioner or **Precision Air Conditioner,** a specialized air conditioner designed for continuous operation, tight control of
temperature and humidity, and use in computer rooms.

Computer Security Institute an organization for information, computer
and network security professionals, see www.gocsi.com for more
information.

Computer Telephony Integration (CTI), technology that enables cooperation and coordination between computer applications and the telephone network. Can be used for many different applications including
CALL CENTER support, INTERACTIVE VOICE RESPONSE, voice mail, fax
automation (in and out), e-mail, etc. and to coordinate applications such
as capturing caller telephone number (via **ANI**, account number (via
IVR) and presenting the voice call and the application information to
the customer service agent.

Computerized Asset Management computer applications for managing physical assets. See **ASSET MANAGER**.

Computerized Maintenance Management System a computer application that supports and organizes maintenance and repair activities for buildings, equipment, vehicles, etc.

Condensing Furnace a type of high-efficiency gas forced-air furnace that uses a second condensing heat exchanger to extract the latent heat from the flue gas.

Conditional Access broadcast industry term for the business processes and systems that manage a customer's ability to access content based on a combination of subscription services, pay per view, parental controls, etc.

Configuration the detailed composition and setup of a system including all the specific devices, the physical arrangement and connections, hardware option settings, **OPERATING SYSTEM** and application software versions, installation parameters, logical names, addresses, etc.

Configuration Management the process of analyzing, planning, and documenting changes to the **CONFIGURATION** of a system. This may include tracking each version of the system and the differences between versions, especially when applied to software.

Configuration Management Database another name for a **CONFIGURATION MANAGEMENT** application that emphasizes the importance of the underlying database.

Consolidation Point (CP), an optional device for interconnecting horizontal cables between the **HORIZONTAL CROSS-CONNECT** and the **TELECOMMUNICATIONS OUTLET** or MUTOA within a **STRUCTURED CABLING SYSTEM**.

Constant Bit Rate video that uses a consistent bit rate while the image quality may vary.

Construction Financial Management Association a professional organization for construction financial professionals and service providers. See www.cfma.org for more information.

Construction Industry Institute a consortium of owners, engineering and construction contractors, and suppliers working to improve the cost effectiveness of the capital facility project life cycle. See www.construction-institute.org for more information.

Construction Sciences Research Foundation conducts research, development, and educational programs to create or enhance tools to improve construction industry communication. See www.csrf.org for more information.

Construction Specifications Canada (CSC) a multi-disciplinary nonprofit association providing education, publications and services for the betterment of the Canadian construction community. For more information see www.csc-dcc.ca.

Construction Specifications Institute an organization that advances the standardization of construction language in building specifications. See www.csinet.org for more information.

Construction Users Roundtable an organization of construction and engineering executives from large consumers of construction services working to improve quality, safety and cost effectiveness in the construction industry. For more information see www.curt.org.

Consumer Electronic Control a channel for remote control functions specified as part of the **HDMI** interface.

Consumer Electronics Association (CEA), a consumer electronics industry association that includes manufacturers, dealers, and installers of audio, home networking/theater, mobile electronics and wireless technology. See www.ce.org for more information.

Contagious Disease or **Communicable Disease**, a disease that may be transmitted from one person or species to another. Contrast to INFECTIOUS DISEASE.

Contaminant Monitoring systems and procedures for detecting biological, chemical, or radioactive contamination of air, water, or other materials.

Content Management processes and software applications for managing digital information in multiple formats. For example, website content management systems enable the development, approval, storage and management of information that may be presented in multiple formats on one or more website.

Continental Automated Buildings Association (CABA), a nonprofit industry organization that promotes the development and understanding of home and building automation, see www.caba.org for more information.

Continuous Acquisition and Lifecycle Support formerly Computer-aided Acquisition and Lifecycle Support, an initiative to develop standards for electronic data interchange, electronic technical documentation, and guidelines for process improvement that was started by the US Department of Defense.

Continuous Availability available at all times, used to imply high AVAILABILITY.

Continuous Operation operational at all times, frequently used to imply high AVAILABILITY.

Contrast the range between the lightest and darkest areas of an image. The contrast control sets the maximum white level for a display screen, see also BRIGHTNESS.

Control Indicating Equipment or **Control and Indicating Equipment**, systems and devices for displaying alarm information and managing alarm systems.

Conventional Air Distribution air distribution using overhead ducts or ceiling plenum, contrast to UNDER FLOOR AIR DISTRIBUTION.

Conventional Fuel traditional energy sources or fossil fuels (petroleum, oil, coal, propane, and natural gas); in some cases nuclear materials such as uranium are also included.

Convergence Retailer retail stores that offer a combination of gas, groceries, hot food, etc.

Cookie or **HTTP Cookie**, a small block text sent by a server to a web browser that is not displayed but may be stored by the browser; the browser will send the cookie or cookies with each request to the server that sent the cookie. Cookies are used for authenticating, tracking, and maintaining specific information about users but they are not programs or viruses.

Cooling Tower equipment used to transfer heat from water or other fluid to the outside air. Wet cooling towers operate on the principle of evaporation, dry cooling towers use convection heat transfer through a surface that divides the working fluid from ambient air.

COP (Coefficient of Performance) a measure of chiller energy efficiency.

Copper Distributed Data Interface a version of the FIBER DISTRIBUTED DATA INTERFACE that uses copper twisted pair cables instead of fiber optics.

Copyleft a form of copyright licensing where the author surrenders some but not all rights in order to make a program free software and to require that all modified and extended versions of that program to be free software as well. For more information see www.gnu.org/licenses/.

CORBA Common Object Request Broker Architecture, an **OMG** standard for communication between software objects running on different computers.

Core 1) a *multi-core* microprocessor contains two or more independent processors in a single package and supports some parallel processing; 2) *magnetic core* memory is an older **RAM** technology replaced by faster semiconductor memories; 3) *'core dump'* is an unformatted copy of memory contents; 4) *building core* the area with the elevators, stairs, bathrooms, shafts and risers typically located at the center of the building; 5) *core network* is the higher speed backbone connections as contrasted to the EDGE NETWORK.

COS (Change of State), a type of event support available in **BACNET** and other systems where instead of continuously polling devices for their state, an application can subscribe to receive a change of state event notification when the state changes. This reduces network traffic and communications processing load.

COSE or **Common Open Software Environment**, a predecessor of the current UNIX standard.

COTS or **Commercial Off-The-Shelf**, parts or systems as contrasted to custom developed or products that are specific to military or government.

Countereavesdropping technology or systems to detect or prevent eavesdropping.

COV (Change of Value), a type of event support available in **BACNET** and other systems where instead of continuously polling devices for their current value, an application can subscribe to receive a change of value event notification when the value changes. This reduces network traffic and communications processing load.

Coverage Area the space served by one BUILDING AUTOMATION SYSTEM device.

CP (Consolidation Point), an optional device for interconnecting horizontal cables between the HORIZONTAL CROSS-CONNECT and the TELECOMMUNICATIONS OUTLET or MUTOA within a STRUCTURED CABLING SYSTEM.

CPFR Collaborative Planning, Forecasting, & Replenishment, a supply chain management technique where manufacturers, customers, and suppliers share information.

CPG Consumer Packaged Goods.

CPI Chemical Process Industry.

CPM COLLABORATIVE PRODUCTION MANAGEMENT, Commercial Property Management, or CRITICAL PATH METHOD.

CPN Collaborative Partner Network.

CPP (Critical Peak Pricing) premium power prices during high demand periods. See DEMAND RESPONSE for more information.

CPTED (Crime Prevention Through Environmental Design), a multi-disciplinary approach to deterring criminal action by combining surveillance, access control, territorial reinforcement and other techniques.

CPU (Central Processing Unit) the part of a computer or microprocessor that does the actual calculating and processing.

CR CARRIAGE RETURN or Condensation Resistance.

CRAC (Computer Room Air Conditioner) or **Precision Air Conditioner**, a specialized air conditioner designed for continuous operation, tight control of temperature and humidity, and use in computer rooms.

CRC or **Cyclic Redundancy Check**, a hash function used to produce a checksum that can be verified to detect errors in data transmission or storage.

CRE Commercial Real-Estate or Corporate Real-Estate.

CREA Certified Real Estate Agent.

Credential a proof of qualification, competence, or clearance associated with a person such as a certificate, license, etc.

CRI Carpet and Rug Institute or Color Rendering Index.

Crime Prevention Through Environmental Design (CPED), a multidisciplinary approach to deterring criminal action by combining surveillance, access control, territorial reinforcement and other techniques.

Critical Infrastructure the facilities that are necessary for a business or other entity to function, may include power, networks, water supply, transportation, etc.

Critical Path Method a process for minimizing project schedules by identifying and optimizing the tasks that control the duration.

Critical Peak Pricing premium power prices during high demand periods. See DEMAND RESPONSE for more information.

CRM (Customer Relationship Management) a computer application for sales, marketing, and customer service (or self-service) support and related record keeping; the application may also include tools for combining customer data from multiple sales and communications channels analytic tools for analyzing customer information.

Cross-Connect a general class of network equipment that enables the termination and connection of cable elements and may support network monitoring and management functions. For example, a network SWITCH, ROUTER, Patch Panel, etc.

Crosstalk any phenomenon by which an electrical signal transmitted on one circuit or channel creates an undesired effect in another circuit or channel.

CRP Capacity Requirements Planning.

CRS (Center for Resource Solutions) a nonprofit organization working to build a robust renewable energy market by increasing demand and supply of renewable resources. For more information see www.resource-solutions.org.

CRT Cathode Ray Tube, display technology used for computers and televisions.

Cryptographic Application Programming Interface (CAPI), an APPLICATION PROGRAMMING INTERFACE to the Microsoft WINDOWS operating system that provides services that enable developers to secure access to Windows-based applications using cryptography (as a way to AUTHENTICATE users) or to enhance the security of stored or transmitted information. See also CRYPTOGRAPHIC SERVICE PROVIDER.

Cryptographic Service Provider (CSP), a special type of Microsoft Windows software library that is used by parts of the CRYPTOGRAPHIC APPLICATION PROGRAMMING INTERFACE to implement encoding and decoding functions based on external security modules or SMART CARD interfaces.

CS (Central Station) the main monitoring point of a security system. See also UL CENTRAL STATION.

CSA Canada Standards Association.

CSAA (Central Station Alarm Association) a trade association for providers, users, bureaus, and other agencies of central station protection services. See www.csaaul.org for more information.

CSC (Construction Specifications Canada) a multi-disciplinary nonprofit association providing education, publications and services for the betterment of the Canadian construction community. For more information see www.csc-dcc.ca.

CSI COMPUTER SECURITY INSTITUTE, CONSTRUCTION SPECIFICATIONS INSTITUTE, or CURRENT SOURCE INVERTER.

CSML Control Systems Modeling Language, a proposed XML schema for modeling control devices.

CSO (Chief Security Officer) high level executive responsible for physical and network security.

CSOX Canadian SARBANES-OXLEY equivalent.

CSP (Cryptographic Service Provider), a special type of Microsoft Windows software library that is used by parts of the CRYPTOGRAPHIC APPLICATION PROGRAMMING INTERFACE to implement encoding and decoding functions based on external security modules or SMART CARD interfaces.

CSR CENTRAL STATION RECEIVER or Customer Service Representative.

CSRF (Construction Sciences Research Foundation) conducts research, development, and educational programs to create or enhance tools to improve construction industry communication. See www.csrf.org for more information.

CSS (Cascading Style Sheets) a language used to describe and control the presentation of documents written in a markup language such as HTML or XHTML; this separates presentation formatting from content and makes it possible to have specialized presentation formats for different device or media types (screen, handheld, print), usage, or user-type. See www.w3.org/Style/CSS/ for more information.

CSV (Comma Separated Values), a file format for representing tabular data as a text file with one line per record and a comma to separate values. Some versions of this format allow quotes around fields that contain reserved characters (such as comma or line endings) or a heading row containing column labels.

CTI Computer Telephony Integration.

CTP Capable-to-Promise or Cordless Telephony Profile.

CTR (Common Telecommunications Room), a shared TELECOMMUNI-CATIONS ROOM.

CTS Compliance Test Specification or CERTIFIED TECHNOLOGY SPECIALIST.

CUA or Common User Access, guidelines for consistent user interface appearance and function originally published by IBM.

CUI (Common User Interface) system or software that provides a consistent user interface to multiple systems or applications from a single window or workstation.

Current Source Inverter an electrical device used to provide variable speed control of motors.

CURT (Construction Users RoundTable) an organization of construction and engineering executives from large consumers of construction services working to improve quality, safety and cost effectiveness in the construction industry. For more information see www.curt.org.

Customer Relationship Management a computer application for sales, marketing, and customer service (or self-service) support and related record keeping; the application may also include tools for combining customer data from multiple sales and communications channels analytic tools for analyzing customer information.

CW Condenser Water.

CWA (Clean Water Act) the primary US law governing water pollution, US Code section 33 subsection 1251.

Cx (Commissioning) the process of preparing systems for use including configuring the systems, start up, performance testing and verification, operator training, and establishing operations and maintenance criteria. The objective of the commissioning process is to ensure that the system functions as designed.

D

D/A Digital-to-Analog.

D-ILA Direct Drive Image Light Amplifier, a form of enhanced LIQUID CRYSTAL ON SILICON technology with a thinner reflective LCD.

DA Discharge Air, see SUPPLY AIR.

DACR (Digital Alarm Communicator Receiver) or **Digital Alarm Communications Receiver,** a system component located at the central station that will receive and display signals from a DIGITAL ALARM COMMUNICATOR TRANSMITTER sent over the public switched telephone network or cellular telephone network.

DACS DIGITAL ALARM COMMUNICATOR SYSTEM, or DIGITAL CROSS-CONNECT SYSTEM.

DACT DIGITAL ALARM COMMUNICATOR TRANSMITTER also known as a **Digital Alarm Communications Transmitter.**

DACUM Developing A Curriculum, a storyboarding process for occupational analysis that creates a picture of a worker's duties, tasks, knowledge, skills, traits, and in some cases tool usage, as the basis for developing educational programs. See www.dacum.org for more information.

Daemon in a computer OPERATING SYSTEM, this term is used to describe a class of programs that run in the background and respond to network requests, hardware activity, or other programs.

DALI (Digital Addressable Lighting Interface) a bidirectional digital protocol for controlling lighting in buildings standardized as part of the IEC 60929 standard for fluorescent lamp ballasts. For more information see www.dali-ag.org.

DAM (Digital Asset Management) a system for storing and organizing the digital representations of intellectual property assets including text, video, audio, etc.

Damper a bladed device used to vary the volume of air passing through an air outlet, inlet, or duct.

DARR (Digital Alarm Radio Receiver) a system component that receives and decodes radio signals and that annunciates the decoded alarm data.

DARS (Digital Alarm Radio System) a system in which alarm signals are transmitted from a DIGITAL ALARM RADIO TRANSMITTER located at a protected premise through a radio channel to a DIGITAL ALARM RADIO RECEIVER at a central station.

DART (Digital Alarm Radio Transmitter) a system component that is connected to, or part of, a DIGITAL ALARM COMMUNICATOR TRANSMITTER and provides radio transmission capability.

DAS Data Acquisition System, or DIRECT ATTACHED STORAGE or DISTRIBUTED ANTENNA SYSTEM.

Dashboard a control panel that provides the information needed to operate a piece of equipment or a system. Dashboards are used in a system and business context to consolidate the display of critical performance measures.

Data a general term for numbers, characters, images, audio, video or other information provided from devices or by users. Data may be analog or digital.

Data Center a facility, or a space within a facility, specially equipped to support critical computer systems with power, cooling, communications, specialized fire suppression, etc. Typically a data center is larger and has more HIGH AVAILABILITY features than a COMPUTER ROOM but the terms are sometimes used interchangeably.

Data Cleansing Data Cleaning, Data Scrubbing, a process for detecting and correcting (or removing) corrupt, inaccurate, or duplicate records from a system or database or preventing their entry into a system. Typically an effort to improve **DATA QUALITY**.

Data Encryption Standard US standard (FIPS 46) method for encrypting data that has been superseded by **AES**.

Data Link Layer (DLL), layer two of the **OSI MODEL** or the **INTERNET PROTOCOL** stack. This layer responds to requests from the network layer and issues service to the physical layer in order to transfer data between adjacent network nodes in a wide area network or between nodes on the same local area network segment.

Data Mining the process of searching large volumes of data for patterns or insight.

Data Quality the completeness, validity, consistency, timeliness and accuracy of data evaluated in terms of the data's fitness its intended uses and the degree to which it correctly represents real-world constructs.

Data Warehouse a computer system designed to archive and organize information over time. Typically data is extracted from production systems and transformed into different formats to support **OLAP**.

Database a structured collection of records stored in a computer so programs retrieve and update the data efficiently. See also **RELATIONAL DATABASE MANAGEMENT SYSTEM, OBJECT DATABASE MANAGEMENT SYSTEM,** and **SCHEMA**.

Database Administrator (DBA), a person or group that maintains a database system, is responsible for the data integrity and performance, and controls the database aspects of **CHANGE MANAGEMENT** for the system.

Database Server 1) a *software* program that provides database management services to other computer programs or computers, as defined by a client-server model or other multi-tier architecture; 2) a *computer* or **SERVER** dedicated to running a database management program.

Daylight Factor a percentage measure reflecting how much of the total light outside the building ends up on an interior working plane or desktop, excluding direct sunlight.

Daylight Glazing windows designed to provide interior illumination and located above eye height (7.5 ft), or the portion of a window more than 7.5 ft above the floor. Also known as DAYLIGHT WINDOW, contrast to VISION GLAZING.

Daylight Harvesting the use of photo sensors to measure available natural light levels and to adjust the amount of artificial lighting added to provide an appropriate lighting level.

Daylight Sensor a device that measures the amount of external light available in an area.

Daylight Window see DAYLIGHT GLAZING.

Daylighting the practice of placing windows, other transparent media, and reflective surfaces so that natural light provides effective internal illumination.

Daylighting Glass see DAYLIGHT GLAZING.

DBA (Database Administrator), a person or group that maintains a database system, is responsible for the data integrity and performance, and controls the database aspects of CHANGE MANAGEMENT for the system.

DBOM Design-Build-Operate-Maintain.

DC Direct Current or Distribution Center.

DCC Digital Command Control.

DCE Distributed Computing Environment, open source distributed processing infrastructure software. See www.opengroup.org/dce/ for more information.

DCOM Distributed Component Object Model, Microsoft technology for distributed processing using **COM**, superseded by .NET.

DCS Digital Cross-Connect System or Distributed Control System.

DCV (Demand Controlled Ventilation) a **HVAC** control system feature where ventilation is adjusted based on actual occupancy and $CO2$ level readings, used to economize cooling for spaces like auditoriums with large variations in occupancy.

DD Design Development or **Device Description.**

DDC Direct Digital Control.

DDE Dynamic Data Exchange, an early technology for information sharing in Microsoft Windows applications.

DDL Device Description Language, a specialized language for expressing **device descriptions.**

DDWG or **Digital Display Working Group,** an industry consortium that developed the **DVI** standard.

DE Digitally Enhanced.

Decision Support System knowledge based computer application designed to support decision making activities or processes. See **Business Intelligence, Data Warehouse, OLAP,** etc.

Decode transforming information that has been encoded or encrypted into its original form or some other usable format. See **Encode** and **Codec.**

Decrypt the process of unscrambling a message that has been **encrypt**ed.

DECT Digital Enhanced Cordless Telephone, an ETSI standard for wireless portable phones (or cordless phones) with an operating range of about 100 meters using frequencies of 1880-1900 MHz in Europe and 1920-1930MHz in the US. For more information see www.dect.org.

Delivery Service Notification an e-mail message containing information about issues related to delivery or non-delivery of a message.

Demand Controlled Ventilation a HVAC control system feature where ventilation is adjusted based on actual occupancy and $CO2$ level readings, used to economize cooling for spaces like auditoriums with large variations in occupancy.

Demand Limiting a system or device to restrict the power draw of a device or building during peak loading conditions.

Demand Management systems that control the load or demand for electric power usually based on dynamic pricing and load information from the power provided by the energy service provider.

Demand Metering electric metering with variable pricing based on the overall balance between demand for power and available supply.

Demand Response technologies to shed load and cut power usage during peak periods. See also DEMAND MANAGEMENT.

Demarcation Point the junction block or other equipment that is considered to be the boundary between the telephone company or other carrier network and the customer premise wiring.

DER (Distributed Energy Resources) a concept where companies that generate heat and power for their own purposes can share excess back into the grid.

DES (Data Encryption Standard) US standard (FIPS 46) method for encrypting data that has been superseded by AES.

Design-Bid-Build a building or system delivery process where design and construction or integration are provided under separate contracts with design services usually provided by a consultant.

Design-Build a building or system delivery process where design and construction or integration are provided a single contract.

Design Development the project phase after conceptual and schematic design and leading into the construction documents phase.

Destination Elevators elevator systems where riders are grouped by floor to minimize travel time and crowding; destination controls can reduce the average journey by 30%. With conventional elevators you get in a car and select your destination; with a destination elevator you specify your destination in the elevator lobby and the control system groups together passengers for optimum transport.

Device Description a formal specification of the data and operating procedures for a device. Depending on the standards and system, there may be a specialized language for writing these descriptions, see **DDL**.

Dew Point the temperature at which a moist air sample cooled at the same pressure would reach water vapor saturation and begin to condense into liquid water, fog, or (if below freezing) solid hoarfrost.

DG Distributed Generation, facilities that can generate heat or electricity for their own needs and optionally sell surplus electrical power back into the grid.

DGI Daylight Glare Index, Disability Glare Index, or Discomfort Glare Index, values range from perceptible (16) to intolerable (28).

DGN CAD file formats supported by Bentley Systems, Intergraph, and others. Originally DGN files were based on Intergraph Standard File Format; in 2000 Bentley Systems created V8 DGN, an updated version with more capabilities and a different internal data structure, see also **OpenDGN**.

DHCP Dynamic Host Configuration Protocol, a network service for automatically assigning IP addresses to devices.

Diagnostic a process or procedure that helps to determine if there is a problem and what the problem may be.

DIAMETER an AUTHENTICATION, AUTHORIZATION, AND ACCOUNTING protocol intended to provide an upgrade from **RADIUS** and standardized as RFC-3588.

DICOM (Digital Imaging and Communications in Medicine) standards for handling, storing, printing, and transmitting medical image information including file formats and a communications protocols. See www.medical.nema.org for more information.

DID (Direct Inward Dialing), a telephone company feature offering that enables customer **PBX** systems (or **VOIP** systems) to route calls to internal extensions automatically. The PBX receives the called number and other signaling information and can make its own routing decisions; this enables a PBX to have fewer connections to the public network than it has extensions (in typical office systems) or more in the case of call centers (see AUTOMATIC CALL DISTRIBUTOR).

Digest a digital summary or hash function computed on a message.

Digital quantitative data represented using discrete values, usually some form of binary number.

Digital Addressable Lighting Interface a bidirectional digital protocol for controlling lighting in buildings standardized as part of the IEC 60929 standard for fluorescent lamp ballasts. For more information see www.dali-ag.org.

Digital Alarm Communicator Receiver or **Digital Alarm Communications Receiver**, a system component located at the central station that will receive and display signals from a DIGITAL ALARM COMMUNICATOR TRANSMITTER sent over the public switched telephone network or cellular telephone network.

Digital Alarm Communicator System a system in which alarm signals are transmitted from a DIGITAL ALARM COMMUNICATOR TRANSMITTER located at the protected premises through the public switched telephone network, cellular telephone network, or other means to a DIGITAL ALARM COMMUNICATOR RECEIVER.

Digital Alarm Communicator Transmitter or **Digital Alarm Communications Transmitter**, a system component located at the protected premises that will contact a DIGITAL ALARM COMMUNICATOR RECEIVER through the public telephone switched network, a cellular telephone system, or other means and transmit the necessary data to identify the protected premises and the change of status. A DACT may be part of the control unit that provides alarm or monitoring functions or interface with a control unit that provides these functions.

Digital Alarm Radio Receiver a system component that receives and decodes radio signals and that annunciates the decoded alarm data.

Digital Alarm Radio System a system in which alarm signals are transmitted from a DIGITAL ALARM RADIO TRANSMITTER located at a protected premise through a radio channel to a DIGITAL ALARM RADIO RECEIVER at a central station.

Digital Alarm Radio Transmitter a system component that is connected to, or part of, a DIGITAL ALARM COMMUNICATOR TRANSMITTER and provides radio transmission capability.

Digital Asset Management a system for storing and organizing the digital representations of intellectual property assets including text, video, audio, etc.

Digital Certificate as part of a PUBLIC KEY INFRASTRUCTURE system the information in a certificate can be used to verify that a public encryption key belongs to an entity with the specified identity (name of a person or an organization, address, etc.) based on the DIGITAL SIGNATURE of a trusted CERTIFICATE AUTHORITY.

Digital Cross-Connect System (DCS), a type of network switching equipment that provides the logical ability to **Cross-Connect** multiple circuits under software control. Some digital cross-connects provide speed and format conversions, automatic switching in the event of link failure and other functions.

Digital Imaging and Communications in Medicine standards for handling, storing, printing, and transmitting medical image information including file formats and a communications protocols. See www.medical.nema.org for more information.

Digital Light Processor a video projection technology that uses tiny mirrors (see **Digital Micromirror Device**) that can be controlled to switch on and off each pixel of the display.

Digital Living Network Alliance (DLNA), a cross-industry collaboration of consumer electronics, computing and mobile device companies developing an interoperability framework and design guidelines for sharing digital media and content services in wired and wireless networks of personal computers, consumer electronics and mobile devices in the home and on the road. For more information see www.dlna.org.

Digital Media any communication media including images, sound, moving pictures, etc. stored in a **digital** format.

Digital Micromirror Device a chip with an array of controllable microscopic mirrors used in **Digital Light Processor** displays.

Digital Rights Management techniques for controlling access to or reproduction of digital information such as music or videos.

Digital Satellite Systems a specific type of satellite dish broadcasting.

Digital Signage electronic screens or signs that can be changed in real-time and used for a variety of purposes including information, advertising, entertainment, and emergency communications. Sign displays may be interactive via **touch screen** or other technology permitting the user to navigate more complex information, send messages, or control other devices.

Digital Signature an encryption technique used to assure the integrity of a message and to positively verify the identity of the message sender.

Digital Subscriber Line technology for high speed data transmission over telephone lines using high frequency signals that allow both DSL and telephone service to be provided on the same wires.

Digital Television (DTV), a system for broadcasting and receiving video and sound by means of digital signals, in contrast to the analog signals used by traditional TV. DTV has the advantage of providing more channels and **HIGH DEFINITION TELEVISION** support but requires replacing or adapting analog television sets. See also **ATSC**.

Digital Theater System a multi-channel digital surround sound format used for commercial/theatrical and consumer grade applications.

Digital Versatile Disc or **Digital Video Disk** (DVD), an optical disk media capable of storing 8.5 **GIGABYTE**s on a 120 mm disk, used primarily for data and video.

Digital Video Monitoring video monitoring using digital systems for communications with cameras, image storage, and feature identification.

Digital Video Recorder a computer system for recording and managing video. System features may include the ability to organize and search recordings in different way, control of playback, remote access, etc.

Digital Videotape a compact video tape format used on DV cassettes or miniDV tapes.

Digital Visual Interface a video interface standard designed to maximize visual display quality by carrying uncompressed video data using a digital protocol in which the desired brightness is transmitted as binary data to LCD displays and digital projectors.

DIN Deutsches Institut für Normung, German Institute for Standardization, the German national standards organization.

Direct/Indirect Lighting an approach to office lighting that provides a mix of light directed down towards the work surface and up toward the ceiling. Lighting fixtures suspended from the ceiling may provide both types of lighting.

Direct Attached Storage disk storage systems that are directly attached to a server or workstation. Contrast to **NAS** and **SAN**.

Direct Inward Dialing (DID), a telephone company feature offering that enables customer **PBX** systems (or **VOIP** systems) to route calls to internal extensions automatically. The PBX receives the called number and other signaling information and can make its own routing decisions; this enables a PBX to have fewer connections to the public network than it has extensions (in typical office systems) or more in the case of call centers (see **AUTOMATIC CALL DISTRIBUTOR**).

Direct Sequence Spread Spectrum a technique for spreading radio data over a very wide band of frequencies in order to minimize the interference caused by noise.

Direct Stream Digital the method of pulse-density modulation encoding of audio signals on digital media used for the Super Audio CD.

Disarm to turn off an alarm or intrusion detection system.

Discharge Monitoring Report a periodic report required under the US **NATIONAL POLLUTANT DISCHARGE ELIMINATION SYSTEM**.

Discovery an automated process for enumerating the devices connected to a network or the elements of a system.

Displacement Ventilation a type of **HVAC** system that introduces air at low velocities at or near the floor in order to cause minimal induction and air mixing. Contrast to **CONVENTIONAL AIR DISTRIBUTION**.

Distributed Antenna System technology for improving indoor coverage and in-building penetration for wireless communications using a network of interior and exterior antennas and active repeater amplifiers. System may be designed to support different radio frequencies and devices including cellular telephone, paging and other data services, or wireless data networks systems.

Distributed Energy Resources a concept where companies that generate heat and power for their own purposes can share excess back into the grid.

Distributed Generation electric power supply from multiple locations and sources including consumer owned cogeneration plants, COMBINED HEATING AND POWER plants, wind farms, etc.

Distributed Processing computer applications involving multiple computers connected together over a network. Computers may be in the same location or geographical disbursed.

Distributor IEC terminology for the equivalent of a CROSS-CONNECT.

District Heating and/or Cooling a system for distributing heat or cooling generated in a central location to multiple residential or commercial buildings. A large district plant can provide higher efficiencies and better pollution control than smaller distributed facilities.

Division 17 proposed addition to the 1995 CSI MASTERFORMAT specification model for telecommunications and technology infrastructure, used as shorthand for these technologies. The 2004 MasterFormat includes several new divisions including Communications (27) and Integrated Automation (25).

Division 25 section of the CSI MasterFormat 2004 specifications for Integrated Automation.

DLL DATA LINK LAYER or **DYNAMICALLY LINKED LIBRARY**.

DLNA (Digital Living Network Alliance), a cross-industry collaboration of consumer electronics, computing and mobile device companies developing an interoperability framework and design guidelines for sharing digital media and content services in wired and wireless networks of personal computers, consumer electronics and mobile devices in the home and on the road. For more information see www.dlna.org.

DLP (Digital Light Processor) a video projection technology that uses tiny mirrors (see DIGITAL MICROMIRROR DEVICE) that can be controlled to switch on and off each pixel of the display.

DM DEMAND MANAGEMENT or Document Management.

DMD (Digital Micromirror Device) a chip with an array of controllable microscopic mirrors used in DIGITAL LIGHT PROCESSOR displays.

DMR (Discharge Monitoring Report) a periodic report required under the US NATIONAL POLLUTANT DISCHARGE ELIMINATION SYSTEM.

DMTF Distributed Management Task Force, an organization that develops and maintains standards for systems management of IT environments in enterprises and the Internet.

DMZ DeMilitarized Zone, a network area in between the public Internet and an organization's more protected internal network.

DNA Microsoft's Distributed iNternet Architecture

DNS DOMAIN NAME SERVER or Domain Name Service.

Document Type Definition a data structure that defines the organization of well formed documents and their contents in a language like SGML, HTML, XHTML or XML.

DoD Department of Defense.

DOE Department of Energy.

DOE-2 building energy analysis software that can predict the energy use and cost for all types of buildings. For more information see www.doe2.com.

Domain Name the identifying label that is registered and assigned under the domain name system (see **DNS**). A domain name may have two or more parts separated by periods and consisting of letters, digits or hyphens; the rightmost part is the top level domain (see **TLD**).

Domain Name Server a specialized network server that translates between a textual **DOMAIN NAME** (see also **URL**) and numeric network address (also called an IP address).

DoS Denial of Service (also, DoS attack), an attack on a computer system that causes a loss of service to users by overloading the resources of the victim system.

Double-Envelope System or **Double-Skin Façade**, a building façade with two layers designed to accommodate additional venting and ventilation practices.

Downlink data transmission from the network, server, or satellite to the user device.

Download to transfer data to the user or client machine. Contrast to UPLOAD.

DP Differential Pressure.

DPI Dots Per Inch, a measure of resolution for printers, scanners, or displays.

DPO Days Payable Outstanding.

DPRS DECT Packet Radio Service, wireless data transmission using **DECT**.

DR DEMAND RESPONSE, Disaster Recovery.

Draft N shorthand for IEEE 802.11n wireless network standard that is currently a draft and subject to change. See **802.11.**

DRAM Dynamic Random Access Memory, a common form of computer **RAM.**

Drill-Down the ability to access supporting data and details behind summary statistics by drilling down to lower levels of detail or the actual source data.

DRM (Digital Rights Management) techniques for controlling access to or reproduction of digital information such as music or videos.

Dropped Ceiling or **Suspended Ceiling,** a secondary ceiling below the bottom of the floor above that can be used conceal piping, wiring, or ductwork and to create a space called the PLENUM.

DRP Distribution Resource Planning.

DSD Direct Store Delivery or DIRECT STREAM DIGITAL.

DSI Digital Signal Interface, a protocol for dimming stage lighting and electrical ballasts, precursor of **DALI.**

DSIRE Database of State Incentives for Renewables and Efficiency, online directory of state, local, utility, and federal incentives that promote renewable energy and energy efficiency. For more information see www.dsireusa.org.

DSL (Digital Subscriber Line) technology for high speed data transmission over telephone lines using high frequency signals that allow both DSL and telephone service to be provided on the same wires.

DSN (Delivery Service Notification) an e-mail message containing information about issues related to delivery or non-delivery of a message.

DSP Digital Signal Processor or Digital Signal Processing, a specially designed chip for signal processing, used in audio, video, radio, etc.

DSS Data Security Standards see **PCI DSS**, **DECISION SUPPORT SYSTEM** or **DIGITAL SATELLITE SYSTEMS**.

DSSS (**Direct Sequence Spread Spectrum**) a technique for spreading radio data over a very wide band of frequencies in order to minimize the interference caused by noise.

DTD (**Document Type Definition**) a data structure that defines the organization of well formed documents and their contents in a language like SGML, HTML, XHTML or XML.

DTE Data Terminal Equipment.

DTMF (**Dual Tone Multi-Frequency**), telephone signaling over the line in the voice band using a push button or matrix keypad, also known as **Touchtone**.

DTS (**Digital Theater System**) a multi-channel digital surround sound format used for commercial/theatrical and consumer grade applications.

DTS-HD Master Audio a digital surround sound format with an unlimited number of channels that is supported on **BLU-RAY DISC** and **HD-DVD**.

DTV **DIGITAL TELEVISION**

Dual Boot the capability to select operating systems when starting a computer. For example, some Macintosh computers can be configured to dual boot Mac OS or Windows.

Dual Core a type of microprocessor that incorporates two central processors (see **CPU** into a single chip.

Dual Tone Multi-Frequency (DTMF), telephone signaling over the line in the voice band using a push button or matrix keypad, also known as **Touchtone**.

Duplex double or two fold: 1) a *building* that contains two residential units; 2) in *data communications* duplex is the ability to transmit data in both directions at once (full duplex) or only one direction at a time (half duplex); 3) *printing or copying* duplexing is printing on both sides of the paper.

DV Digital Videotape or Displacement Ventilation.

DVD (Digital Versatile Disc) or **Digital Video Disk (DVD)**, an optical disk media capable of storing 8.5 Gigabytes on a 120 mm disk, used primarily for data and video.

DVI (Digital Visual Interface) a video interface standard designed to maximize visual display quality by carrying uncompressed video data using a digital protocol in which the desired brightness is transmitted as binary data to LCD displays and digital projectors.

DVR (Digital Video Recorder) a computer system for recording and managing video. System features may include the ability to organize and search recordings in different way, control of playback, remote access, etc.

DW Data Warehouse, a specialized database used to store and analyze historical information extracted from operational databases. Typically this data is extracted, transformed and then loaded into a specially designed subject oriented database.

DWDM Dense Wavelength Division Multiplexing, systems with 16 or more channels.

DWG the proprietary file format developed by Autodesk for storing design data created using AutoCAD and related products. See also OpenDWG.

Dx Diagnosis.

DX Direct Expansion.

DXF AutoCAD Drawing Interchange Format or Drawing Exchange Format, **CAD** data file formats developed and published by Autodesk to enable data interoperability between AutoCAD and other programs.

Dynamically Linked Library or **Dynamic Library** (DLL), a software library that is incorporated at runtime.

E

E&M or **Ear-and-Mouth,** telephone interface connections to speaker and microphone.

E1 or **E-1,** a data circuit with a speed of 2.048 Mbit/s.

e-Waste electronics equipment classified as hazardous waste due to the presence of lead or mercury and subject to state and federal disposal restrictions.

EA ENERGY AND ATMOSPHERE, Engineering and Architectural firm, or Exhaust Air.

EAI (Enterprise Application Integration), systems and technologies for integrating applications within an organization. See also INTEGRATION and for example, www.icmembers.org.

EAL (Evaluation Assurance Level), a numerical grade from 1 to 7 assigned to a computer product or system following the completion of a COMMON CRITERIA security evaluation based on the extent and type of testing. EAL should be considered together with the PROTECTION PROFILE to determine if a product's fitness for a given application.

EAM Enterprise Asset Management.

EAN EUROPEAN ARTICLE NUMBER or Event Alerting and Notification, see EVENT MANAGEMENT.

EAP Extensible Authentication Protocol, RFC-3748 universal authentication framework, used in wireless networks and point-to-point connections.

Early Warning Smoke Detector technology to detect fire prior to the time that it becomes threatening to building occupants, generally at the time that smoke is visible.

EAS Electronic Article Surveillance, a technology designed to prevent shoplifting or theft from libraries using **RFID** or other tags fixed to merchandise or books.

EB (Exabyte) one quintillion, or ten to the eighteenth power, bytes (characters) of storage. See Fig. 1 on page 279.

EC Electrical Contractor or **Electrochromic**.

ECI Energy Conservation Initiative, see **Energy Conservation Measure**, or Evangelical Climate Initiative.

ECM **Energy Conservation Measure** or **Enterprise Content Management**.

ECMAScript see **JavaScript**.

ECN Engineering Change Notice.

Economizer a type of **HVAC** device designed to save energy by using outside air as a means of cooling the indoor space. Air-side economizers cool by bringing in outside air; water-side economizers use cooling towers or heat exchangers.

Ecoprofile the outcome of an **life cycle assessment** study defining the impact on climate change, toxicity, fossil fuel depletion, water resources, etc. of a building or a material.

ECR Efficient Consumer Response, European grocery sector supply chain and operational standards initiative, see www.ecrnet.org for details.

eCRM Electronic Customer Relationship Management

EDA (Equipment Distribution Area), the location of equipment cabinets and racks in a data center, this space typically also includes **Patch Panels** for connecting equipment.

EDGE Enhanced Data rates for GSM Evolution or Enhanced GPRS (EGPRS), is a broadband data service using **GSM** networks.

Edge Device a general class of networking routers, switches and other devices that connect the **Edge Network** to the faster backbone or **core** network such as an enterprise, carrier, or service provider network.

Edge Network 1) *networking* a general term for networks on the periphery of the **core** network with the connections to users and devices, in a structured cabling system this would include the Horizontal Cabling and Work Area components; 2) a *wireless network* offering **EDGE** service.

EDI Electronic Data Interchange.

EDID (Extended Display Identification Data) a **VESA** standard data structure provided by a computer display to describe its capabilities to a graphics card.

EDM Electronic Data Management, Electrical Discharge Machine, or Electronic Document Management.

EE Electrical Engineer or Energy Efficiency.

EEM Energy Efficiency Measure, a change in design, technology or operation intended to reduce energy requirements.

EEPROM Electronically Erasable Programmable Read Only Memory.

EF (Entrance Facilities), the cables, connecting hardware, protection devices and other equipment required to connect the carrier's outside plant facilities to the premises cabling. See also **Entrance Room**.

Efficiency Valuation Organization a non-profit organization dedicated to creating measurement and verification (**M&V**) tools in support of energy efficiency, renewable energy, and water efficiency. For more information see www.evo-world.org.

eFPM E-Fulfillment Process Management.

eFS E-Fulfillment Solutions.

EFT Electronics Funds Transfer.

EHS Environmental Health and Safety.

EIA Electronics Industries Alliance or Energy Information Administration.

EIA-232 (formerly RS-232) serial interface specification replaced by **TIA-232**.

EIA-485 (formerly RS-485) a physical layer specification for a two-wire, half-duplex, multipoint serial connection.

EIA-852 Tunneling Specification, the official standard for **LonTalk/IP**.

EIB European Installation Bus, see **KNX**.

EICTA European Information & Communications Technology Industry Association.

EIFS Exterior Insulation and Finish System, a multilayer exterior wall construction that includes rigid insulation boards mounted outside the wall sheathing and a waterproof elastomeric exterior skin.

eIS E-Integration Solutions

EIS (Energy Information System) applications for tracking and managing the use of electricity and other forms of energy.

EL Electroluminescence.

Electric Service Provider (ESP), an ENERGY SERVICE PROVIDER offering electric power in a competitive market.

Electro Magnetic Interference or Electromagnetic Interference, noise interruption or disruption of function caused by electromagnetic radiation. Technically electromagnetic interference is a more general term than RADIO FREQUENCY INTERFERENCE but they are typically used interchangeably.

Electrochromic tungsten oxide based materials used to electrically control the amount of light and heat allowed to pass through SMART GLASS.

Electronic Product Information Cooperation or **European Product Information Co-operation,** a standard construction product classification system that is being incorporated into OMNICLASS. For more information see www.epicproducts.org.

Electronics Industries Alliance organization that, among other things, facilitates the process of developing standards and submits them to ANSI for adoption as American National Standards and international standards.

Electrostatic Discharge a static shock or spark that may damage equipment.

EMCS Energy Management and Control System or Energy Monitoring and Control System. See ENERGY MANAGEMENT SYSTEM.

Emergency Power Off or a 'disconnecting means' mechanism for turning off power to a piece of equipment or an entire facility in the event of a fire or accident. This includes the ability to override all secondary power sources or UNINTERRUPTIBLE POWER SUPPLY backup power.

Emergency Shut Down a system for removing power and shutting down a system or process. See also EMERGENCY POWER OFF.

EMI (Electro Magnetic Interference) or Electromagnetic Interference, noise interruption or disruption of function caused by electromagnetic radiation. Technically electromagnetic interference is a more general term than RADIO FREQUENCY INTERFERENCE but they are typically used interchangeably.

EMPP (Equal Marginal Performance Principle) a way of looking at systems composed of variable speed power-modulating components, such as fans, chillers, and pumps, optimizing the design so that under peak load adding the same marginal power to any system component will yield the same increase in cooling output and controlling the system using algorithms that ensures optimal performance at all possible loading levels.

EMS Emergency Medical Services or ENERGY MANAGEMENT SYSTEM.

EMV Payment card specifications developed by Europay, Mastercard International and Visa. For more information see www.emvco.com.

Encapsulate to wrap or hide information or details. A software object encapsulates all the details of its data and processing logic and only exposes certain interface functions or methods. One communications network protocol may be encapsulated or wrapped inside another protocol.

Encode to convert information into a particular coding system or format. Contrast to DECODE, see also CODEC.

Encrypt the process of scrambling or obscuring data so that it can't be read by an unauthorized party.

Encryption Key data used to ENCRYPT and DECRYPT a message.

Energy and Atmosphere (EA), LEED terminology for all things related to energy usage/efficiency, GREENHOUSE GAS emissions and ozone protection.

Energy Conservation Measure general term for changes made in order to minimize energy requirements.

Energy Efficient Lighting lamps and light fixtures that use at least 2/3 less energy than conventional incandescent lights.

Energy Harvesting any technology designed to capture energy from the environment and convert it into usable power, typically as electricity. For example, a SOLAR PANEL converts light, a piezoelectric device converts motion, etc.

Energy Information Administration part of the US Department of Energy. See www.eia.doe.gov for more information.

Energy Information System applications for tracking and managing the use of electricity and other forms of energy.

Energy Management System an application for measuring and tracking electrical energy usage (see SUB-METERING) that may also support DEMAND MANAGEMENT and LOAD SHEDDING.

Energy Service Company (ESCO), a business providing energy management services to an energy user either on a fee for service basis or under a performance based contract paid by energy cost savings.

Energy Service Provider (ESP), 1) a *vendor* of energy provided from conventional or alternative sources; 2) a *provider* of energy management services, see ENERGY SERVICE COMPANY.

Energy Usage Intensity (EUI) is an energy measure used by the US Department of Energy and others to rate facilities based on total annual energy usage per square foot.

EnergyPlus simulation software for modeling heating, cooling, lighting, ventilating, and other energy flows in buildings. For more information see www.eere.energy.gov/buildings/energyplus/.

EnergyStar a US government program to help businesses and individuals protect the environment through superior energy efficiency. For more information see www.energystar.gov.

EnOcean very low power WIRELESS SENSOR NETWORK technology that uses the ISM BAND (868.3 MHz) and can be used in battery-less devices that rely on ENERGY HARVESTING for power.

Enterprise any business or organization, typically used to mean the entire organization covering multiple business units and locations.

Enterprise Application Integration (EAI), systems and technologies for integrating applications within an organization. See also INTEGRATION and for example, www.icmembers.org.

Enterprise Content Management processes and software applications for CONTENT MANAGEMENT functions in support of multiple applications and business units.

Enterprise Resource Planning applications that automate many business processes associated with the operations or production aspects of a company.

Enterprise Service Bus (ESB), a software architecture where a standards-based messaging engine (the bus) provides the foundation for multiple event-driven applications running on multiple computers, possibly at multiple locations. Frequently used for ENTERPRISE APPLICATION INTEGRATION and may use WEB SERVICES, MQ or MIDDLEWARE to create the bus.

Enthalpy a measure of the heat content (internal energy) of a specific quantity of material at a given temperature in units of energy per unit of mass. In North America enthalpy is expressed in BTU/lb; the SI units are kJ/kg.

Entrance Facilities (EF), the cables, connecting hardware, protection devices and other equipment required to connect the carrier's outside plant facilities to the premises cabling. See also ENTRANCE ROOM.

Entrance Room a telecommunications space used for access provider equipment, DEMARCATION POINTs, and the interface to the campus cabling system.

Environmental Tobacco Smoke indoor air pollution from tobacco use, also known as second hand smoke.

EOC Energy Operations Center.

EOL Resistor an end-of-line resistor or relay is used to monitor the health of an alarm or fire detector circuit. The panel or supervisory system monitoring the wire expects a certain resistance value; a change indicates trouble with the wiring.

EOS Economy of Scale.

EPA Environmental Protection Agency.

EPAct Energy Policy Act of 2005, major US energy law, see www.ferc.gov for more information.

EPC Electronic Product Code or Engineering, Procurement, and Construction.

EPIC Electronic Privacy Information Center or ELECTRONIC PRODUCT INFORMATION COOPERATION.

EPM Enterprise Production Management

EPO (Emergency Power Off) or a 'disconnecting means' mechanism for turning off power to a piece of equipment or an entire facility in the event of a fire or accident. This includes the ability to override all secondary power sources or UNINTERRUPTIBLE POWER SUPPLY backup power.

ePS E-Procurement Solutions

EPS Enterprise Production System.

Equal Marginal Performance Principle a way of looking at systems composed of variable speed power-modulating components, such as fans, chillers, and pumps, optimizing the design so that under peak load adding the same marginal power to any system component will yield the same increase in cooling output and controlling the system using algorithms that ensures optimal performance at all possible loading levels.

eQUEST a Windows implementation of **DOE-2**.

Equipment Distribution Area (EDA), the location of equipment cabinets and racks in a data center, this space typically also includes PATCH PANELS for connecting equipment.

Equipment Room (ER), TIA-862 terminology for a MECHANICAL ROOM where BUILDING AUTOMATION SYSTEM controllers are located.

ER (Equipment Room), TIA-862 terminology for a MECHANICAL ROOM where BUILDING AUTOMATION SYSTEM controllers are located.

ERP (Enterprise Resource Planning) applications that automate many business processes associated with the operations or production aspects of a company.

ESB (Enterprise Service Bus), a software architecture where a standards-based messaging engine (the bus) provides the foundation for multiple event-driven applications running on multiple computers, possibly at multiple locations. Frequently used for ENTERPRISE APPLICATION INTEGRATION and may use WEB SERVICES, MQ or MIDDLEWARE to create the bus.

ESCO (Energy Service Company), a business providing energy management services to an energy user either on a fee for service basis or under a performance based contract paid by energy cost savings.

ESD ELECTROSTATIC DISCHARGE or EMERGENCY SHUT DOWN.

ESP ELECTRIC SERVICE PROVIDER or ENERGY SERVICE PROVIDER.

EtherCAT Ethernet for Control Automation Technology, an open standard for Ethernet-based FIELDBUS standardized as IEC/PAS 62407. See www.ethercat.org for more information.

Ethernet a computer networking technology originally developed for local area networks and extended to include wireless and metropolitan area networks. Ethernet standards are defined in the IEEE **802.3** and **802.11** series.

EtherNet/IP Ethernet/Industrial Protocol, standard for COMMON INDUSTRIAL PROTOCOL over Ethernet. See www.odva.org for more information.

Ethernet Hub or **Ethernet Repeater,** a device that connects together multiple Ethernet devices and making them act as a single network segment. Hubs have largely been replaced by Ethernet SWITCHes which provide isolation between segments and the ability to monitor each segment.

ETL (Extract, Transform and Load), a process that is used to move information from an operational data system or ONLINE TRANSACTION PROCESSING SYSTEM to a DATA WAREHOUSE or another operations system; this process involves Extracting or copying data, Transforming the format, coding or structure to meet the requirements of the target system, and Loading the data into the target system. This process may be continuous (with a message queue) or periodic; the timing for periodic updates may range from multiple times per day to monthly or quarterly.

ETS (Environmental Tobacco Smoke) indoor air pollution from tobacco use, also known as second hand smoke.

EU European Union.

EUI (Energy Usage Intensity) is an energy measure used by the US Department of Energy and others to rate facilities based on total annual energy usage per square foot.

European Article Number (EAN), a barcode standard which is a superset of **UPC** and also administered by GS1. See www.gs1.org for more information.

EV-DO EVDO, or **EV,** Evolution-Data Optimized, a broadband data service using **CDMA2000** wireless networks.

EV SSL Extended Validation Secure-Sockets-Layer, a type of enhanced security certificate where the certificate authority affirms that private data are being encrypted and that the identity of the business operating the site has been confirmed. See also **SSL.**

Evaluation Assurance Level (EAL), a numerical grade from 1 to 7 assigned to a computer product or system following the completion of a COMMON CRITERIA security evaluation based on the extent and type of testing. EAL should be considered together with the PROTECTION PROFILE to determine if a product's fitness for a given application.

EVDO EVolution Data Only, data networking technology using cellular telephone networks.

Event Management 1) *social or business event* organizing and administration and supporting business processes and applications; 2) ALARM and ALERT management, processes and applications for tracking, diagnosing, and correcting things that happen in a system, building, or network.

EVO (Efficiency Valuation Organization) a non-profit organization dedicated to creating measurement and verification (M&V) tools in support of energy efficiency, renewable energy, and water efficiency. For more information see www.evo-world.org.

EWSD (Early Warning Smoke Detector) technology to detect fire prior to the time that it becomes threatening to building occupants, generally at the time that smoke is visible.

Exabyte one quintillion, or ten to the eighteenth power, bytes (characters) of storage. See Fig. 1 on page 279.

Exchange 1) a Microsoft server product that supports e-mail and other collaboration features; 2) an old term for a telephone CENTRAL OFFICE.

Exploit a general term for a technique used to take advantage of a computer security vulnerability including software, data, or a sequence of commands.

EXPRESS a language for unambiguous lexical product data standardized as ISO 10303-11. This standard defines data types and constraints on instances of the data types.

EXPRESS-G a graphical representation for a subset of the constructs in the **EXPRESS** language standardized as ISO 10303-11.

Extended Display Identification Data a **VESA** standard data structure provided by a computer display to describe its capabilities to a graphics card.

Extensible Messaging and Presence Protocol an open XML-based protocol for near-real-time instant messaging and PRESENCE INFORMATION. Standard was based on Jabber technology and formalized as RFC 3920 and RFC 3921.

Extract, Transform and Load (ETL), a process that is used to move information from an operational data system or ONLINE TRANSACTION PROCESSING SYSTEM to a DATA WAREHOUSE or another operations system; this process involves Extracting or copying data, Transforming the format, coding or structure to meet the requirements of the target system, and Loading the data into the target system. This process may be continuous (with a message queue) or periodic; the timing for periodic updates may range from multiple times per day to monthly or quarterly.

F

FA Factory Automation or FIRE ALARM.

Facilities Management or **Facility Management,** the practice of coordinating the physical workplace with the people and work of the organization; integrating principles of business administration, architecture, behavioral and engineering sciences.

Facility Information Council a NATIONAL INSTITUTE OF BUILDING SCIENCES council that supports the development, standardization, and integration of computer technologies and software to ensure the improved performance of the entire life cycle of facilities. For more information see www.facilityinformationcouncil.org.

Facility Life Cycle the CONSTRUCTION SPECIFICATIONS INSTITUTE defines the life cycle as the stages of a project leading to a completed facility (project conception, project delivery, design, construction documentation, procurement, construction and facility management), its use and management until its eventual modification, deconstruction, demolition or adaptive reuse.

Facility Management System see COMPUTER AIDED FACILITIES MANAGEMENT.

Facility Operations Portal a PORTAL designed to give operations and maintenance staff access to the information they need based on their role. This may include access to building automation systems, alarm management, work orders, occupant information, building models and documentation, procedures, etc.

FACP (Fire Alarm Control Panel) or **Fire Alarm Control Unit,** one element of a FIRE ALARM system for detecting and reporting occurrences of fires within a building based on inputs from automatic and manual fire alarm devices. The system may supply power to detection devices, transponders, off-premises transmitters, or NOTIFICATION APPLIANCEs and transfer of condition to relays or devices connected to the control unit including FIRE SUPPRESSION SYSTEMS, activating, closing fire doors, unlocking exit doors, and automatically activating the SMOKE CONTROL SYSTEM in some situations. The fire alarm control unit can be a local fire alarm control unit for part of a building or a master control unit.

FACU Fire Alarm Control Unit, see **Fire Alarm Control Panel**.

Façade the exterior of a building or one side of the exterior of a building, especially the front. See also **Active Façade**.

FC (Fibre Channel) high speed storage networking standard that uses copper or fiber optic cables.

FCA Facility Condition Assessment.

FCC Federal Communications Commission, US government agency that regulates interstate and international communications by radio, television, wire, satellite and cable. See www.fcc.gov for more information.

FCIP or **FC/IP, Fibre Channel** over **IP** protocol, a **storage area network** technology that facilitates data sharing between distributed facilities.

FCS Field Control System

FCU Fan Coil Unit.

FD Floor Diffuser.

FDA Food & Drug Administration, the US government agency that regulates the safety of the food supply, medicine, and medical equipment.

FDC Factory Data Collection, systems for machine monitoring and production management.

FDDI (Fiber Distributed Data Interface) standard for local area network data transmission using dual attached counter rotating token rings, formalized as ANSI X3T9.

FEC Field Equipment Controller.

FEMP the Federal Energy Management Program of the US Department of Energy.

Fenestration 1) design and positioning of windows; 2) any structural opening in a building envelope; 3) windows, doors and skylights.

Feng Shui the Chinese art of managing the movement of chi—the life force—within a space to achieve harmony with the environment.

FF Fieldbus Foundation, non-profit education and promotion organization.

FF&E (Furniture, Fixtures and Equipment) items that are not considered to be permanently attached to a structure and may be accounted for separately from the base building capital expense.

FFD (Full Function Device) a complex device in a building automation system such as a variable air volume control or a plant controller. Contrast to **RFD**.

Fiber Distributed Data Interface standard for local area network data transmission using dual attached counter rotating token rings, formalized as ANSI X3T9.

Fiber To The Curb a telecommunications network where fiber-optic cables run to a platform that serves several customers.

Fiber To The Premise fiber optic communications network that extends to the customer's building.

Fiber to the Telecom Enclosure a structured cabling system technology that extends the fiber backbone from the equipment room, through the riser and telecom room, directly to a Telecommunications Enclosure (TE) serving the work area.

Fibre Channel high speed storage networking standard that uses copper or fiber optic cables.

FIC (Facility Information Council) a National Institute of Building Sciences council that supports the development, standardization, and integration of computer technologies and software to ensure the improved performance of the entire life cycle of facilities. For more information see www.facilityinformationcouncil.org.

Fieldbus a networking system for real-time distributed industrial control, standardized as IEC 61158.

FIFO (First In First Out), a production, inventory management or accounting method where the oldest item is used first, contrast to **LIFO**.

File Transport Protocol a protocol for copying or accessing files over **TCP/IP** networks.

FINREAD FINancial Transactional IC Card READer, a European **Smart Card** consortium and standards development organization. For more information see www.finread.com.

FIP Factory Instrumentation Protocol

FIPS 201 or **Federal Information Processing Standard 201,** a US government standard that specifies **Personal Identity Verification** requirements for employees and contractors with access to Federal facilities and information systems. Agencies are using combinations of **Smart Card, Biometric** information and **Public Key Infrastructure** systems.

Fire Alarm systems for detecting fire, alerting occupants and first responders, and coordinating fire suppression and smoke control actions.

Fire Alarm Control Panel or **Fire Alarm Control Unit,** one element of a **Fire Alarm** system for detecting and reporting occurrences of fires within a building based on inputs from automatic and manual fire alarm devices. The system may supply power to detection devices, transponders, off-premises transmitters, or **Notification Appliances** and transfer of condition to relays or devices connected to the control unit including **Fire Suppression Systems,** activating, closing fire doors, unlocking exit doors, and automatically activating the **Smoke Control System** in some situations. The fire alarm control unit can be a local fire alarm control unit for part of a building or a master control unit.

Fire Alarm Notification System automatic and manual communications system for use in the event of fire and other emergencies. The system may control **NOTIFICATION APPLIANCE**s, include pre-recorded audio messages and public address capabilities, and **DIGITAL SIGNAGE**.

Fire Alarm Signaling System devices for automatically communicating fire alarm information, see **DIGITAL ALARM COMMUNICATOR SYSTEM** and **DIGITAL ALARM RADIO SYSTEM**.

Fire and Life Safety or **Fire, Life and Safety**, systems for detecting fire and protecting life and property through sensors, pull stations, alarm, fire doors, smoke dampers, ventilation controls, etc.

Fire Batt an insulating material used for **FIRESTOP**s and **FIREWALL** construction.

Fire Dialer automatic dialer as part of a fire signaling system, see **DIGITAL ALARM COMMUNICATOR TRANSMITTER**.

Fire Paint see **INTUMESCENT PAINT**.

Fire Protective Signaling System see **FIRE ALARM SIGNALING SYSTEM**.

Fire Rating a classification indicating how long a structure or component may be expected to withstand a standardized fire test. This classification does not necessarily reflect performance of components in an actual fire.

Fire Suppression Systems automatic systems designed to protect people and property in the event of fire. Depending on the types of fire hazards and property involved this may include sprinkler systems (wet, dry, pre-action, and deluge), gaseous agents, wet or dry chemical agents.

Firefighters Smoke Control Station (FSCS), a device that enables firefighters to manually control the **SMOKE CONTROL SYSTEM**, overriding automatic control of system components. The FSCS provides a graphical representation of the building, smoke control zones, and smoke control equipment including the mode of each zone and equipment status.

Firestop or **Firestopping,** specialized materials and processes used to seal openings in fire-resistance rated wall or floor assemblies in order to restore the fire-resistance rating. Firestop is used for electrical, mechanical, and structural through-penetrations and junctions between rated assemblies.

Firewall 1) in *construction* a firewall, **fire-rated wall,** or a **rated wall** is a wall that is reinforced to resist burning; 2) in *networking* a firewall is a security device or software that restricts access to a computer or subnetwork by selectively blocking messages that could be harmful.

FireWire IEEE 1394, Sony i.Link, a serial BUS interface standard used for computers and video equipment. FireWire 400 can transfer data at 100, 200, or 400 Mbit/s.

First Costs the expenses incurred for designing, building, and commissioning a new building or CAPEX. Contrast to OPEX, LIFE CYCLE COST or TOTAL COST OF OWNERSHIP.

First In First Out (FIFO), a production, inventory management or accounting method where the oldest item is used first, contrast to **LIFO.**

First Responder any trained emergency worker expected to be among the first to arrive on the scene of an attack, emergency event, or other incident; first responders include firefighters, security guards, and police officers. In fire terminology the First Responder is the first fire command to arrive. Emergency Medical Services use this term to mean the first individual who arrives at the scene regardless of the individual's type of credential and offer First Responder training for nonprofessionals.

FISMA Federal Information Security Management Act, a 2002 law that bolsters government computer and network security by mandating yearly audits.

Five Nines 99.999% AVAILABILITY for a system or service. See Fig. 4 on page 281.

Flash 1) a type of semiconductor *memory* that is rewritable and holds its contents without power; 2) a family of multimedia *file formats* and software tools originally Macromedia and now offered by Adobe.

Flash Drive a device that uses **FLASH** semiconductor memory to provide functionality similar to a disk drive. Typically these thumb sized devices plug into a **USB** port and do not require an external power supply.

Floor Distributor IEC terminology for a **HORIZONTAL CROSS-CONNECT**.

Floor Plate another term for floor layout.

Flow Switch see **SPRINKLER FLOW SWITCH**.

FLS (Fire and Life Safety) or **Fire, Life and Safety,** systems for detecting fire and protecting life and property through sensors, pull stations, alarm, fire doors, smoke dampers, ventilation controls, etc.

FM FACILITIES MANAGEMENT or **FREQUENCY MODULATION**.

FM-200 brand name for Heptafluoropropane fire suppressant gas.

FMP Flexible Manufacturing Plant.

FMS FACILITY MANAGEMENT SYSTEM, Factory Management System, or Factory Message Specification.

Fob System similar to a **SMART CARD** system but based on devices with a key fob form-factor, typically containing a **PASSIVE RFID** device.

Follow-the-Sun a business model where operational responsibility is passed between centers in different geographic locations and time zones around the world.

FOLS Fiber Optic LAN Section, a **TIA** working group that provides education and develops standards for optical applications in customer-owned networks.

Forest Stewardship Council an international membership organization working to find solutions to the problems created by bad forestry practices and to reward good forest management. For more information see www.fsc.org.

Format 1) a specific way of organizing and representing information; 2) the process of transforming information for display or other usage.

Fortran Formula Translation, a venerable computer programming language.

FPS Frames Per Second, in video images.

Frame 1) in *construction* a frame is a structural system that supports or surrounds other components; 2) in *video* or *film* a frame is a single image of a motion picture; 3) in *digital networking* a frame is a data packet of fixed or variable length; 4) in *telephony* or *cabling* a frame or a distribution frame is a type of equipment where cables are connected; 5) in *web pages* HTML frames provide a way to display multiple documents in different areas of the screen.

Free Space Optics 'wireless' optical communications without optical fiber or other dedicated media, may use infrared, laser or other technologies.

Freeze a CHANGE MANAGEMENT practice that prohibits or minimizes changes to IT systems during critical time periods. For example, retailers may freeze systems during the Christmas shopping season and manufacturing companies may freeze systems prior to the end of their fiscal year.

Frequency Modulation a signaling technique where information is conveyed using changes in the frequency information. FM radio uses frequency modulation signaling in the 87.5 to 108.0 MHz band (76 to 90 MHz in Japan).

Frit Glass or **Fritted Glass,** glass with a ceramic-enamel coating to control solar radiation.

Front-of-House the parts of a facility that are open to the public and patrons; contrast to **Back-of-House**.

FRU Field Replaceable Unit.

FSC (Forest Stewardship Council) an international membership organization working to find solutions to the problems created by bad forestry practices and to reward good forest management. For more information see www.fsc.org.

FSCS (Firefighters Smoke Control Station), a device that enables firefighters to manually control the **smoke control system**, overriding automatic control of system components. The FSCS provides a graphical representation of the building, smoke control zones, and smoke control equipment including the mode of each zone and equipment status.

FSK Frequency Shift Keying, a way to encode data using different frequencies.

FSO (Free Space Optics) 'wireless' optical communications without optical fiber or other dedicated media, may use infrared, laser or other technologies.

FT-IR or FTIR Fourier Transform InfraRed spectroscopy, a measurement technique used to identify substances and measure concentrations.

FTC Federal Trade Commission, a US government agency.

FTE Full-Time Equivalent workers.

FTP (File Transport Protocol) a protocol for copying or accessing files over **TCP/IP** networks.

FTTC **Fiber To The Curb** or Fiber To The Cabinet.

FTTE (Fiber to The Telecom Enclosure) a structured cabling system technology that extends the fiber backbone from the equipment room, through the riser and telecom room, directly to a **Telecommunications Enclosure** (TE) serving the work area.

FTTP (Fiber To The Premise) fiber optic communications network that extends to the customer's building.

Fuel Shifting the ability to switch between different fuels as sources of energy for heating or power generation.

Full Function Device a complex device in a building automation system such as a variable air volume control or a plant controller. Contrast to **RFD**.

Fume Hood laboratory equipment designed to limit exposure to airborne hazards including toxic chemicals and biological agents by drawing air away from the user. There are many types of hoods designed for different hazards including externally ducted or re-circulating hoods equipped with different types of filters or washing and scrubbing systems.

Furniture, Fixtures and Equipment items that are not considered to be permanently attached to a structure and may be accounted for separately from the base building capital expense.

Fusible Link is a device made from a fusible alloy that is designed to melt at a specific temperature. Fusible links are used as triggering devices in fire sprinklers, mechanical automatic door releases, and other applications.

Fuzzy Logic a form of automated reasoning that uses membership functions and ranges instead of precise values that can be applied to control problems that involve sub-ranges of a continuous variable. Not to be confused with fuzzy thinking.

G

Galleria a covered passageway or indoor court containing a variety of shops or businesses.

Gantt Chart a project management tool based on a horizontal bar graph that shows the plan and progress for each task or group of tasks based on start date and duration.

Gas Chromatograph an instrument for separating chemicals in a complex sample and measuring the components.

Gas Monitoring technology for monitoring gas leakage, inventory, usage, etc.

Gas Service Provider (GSP), an **ENERGY SERVICE PROVIDER** offering natural gas in a competitive market.

Gateway an adapter or converter that provides an interface between different systems or protocols.

Gateway Load Balancing Protocol (GLBP), a Cisco proprietary protocol that adds basic load balancing functionality to existing redundant router protocols.

GB (Gigabyte) one billion, or ten to the ninth power, bytes (characters) of storage. See Fig. 1 on page 279.

GBA Gross Building Area.

GBC see **USGBC**.

gbXML the Green Building XML schema, was developed to facilitate the transfer of **CAD** building information models between design models, a variety of engineering analysis tools, and other types of models. See www.gbxml.org for more information.

GC G**AS** C**HROMATOGRAPH** or General Contractor.

GCS (**Generic Cabling System**) standard cabling designed to support multiple applications, see S**TRUCTURED** C**ABLING** S**YSTEM** and TIA-568 for the commercial versions.

GE G**IGABIT** E**THERNET**.

General Motion Control technology used for manufacturing automation including general purpose motion controller systems and devices like servo and stepper drives or motors.

General Public License or **GNU General Public License**, a widely-used O**PEN** S**OURCE** software license developed for the GNU project. The GPL grants the rights of the free software definition and uses C**OPYLEFT** to ensure these freedoms are preserved, even when the work is changed or extended; see www.gnu.org/licenses/ for more information.

Generic Cabling System standard cabling designed to support multiple applications, see S**TRUCTURED** C**ABLING** S**YSTEM** and TIA-568 for the commercial versions.

Geographic Information System a system for storing, manipulating, and displaying information linked to geographic locations.

Geolocation process or software for determining the geographic location of a device based on its Internet Protocol address, MAC address, hardware embedded identification number, or other information.

Geospatial computing or information services using location data.

GHG (**Greenhouse Gas**) a gas that contributes to the greenhouse effect such as carbon dioxide, methane, nitrous oxide, and ozone.

GI Glare Index.

GIF (**Graphics Interchange Format**) a bit-mapped image format with a limited color range that is widely used for simple images and animations.

Gigabit data transmission speed measured in billions of bits per second, one gigabit per second is the same as 1,000 megabits per second. See Fig. 2 on page 279.

Gigabyte one billion, or ten to the ninth power, bytes (characters) of storage. See Fig. 1 on page 279.

GIS (Geographic Information System) a system for storing, manipulating, and displaying information linked to geographic locations.

GL General Ledger.

GLA Gross Leasable Area.

Glass Break Detector or **Glass Break Sensor,** devices that detect breaking glass based on sound, vibration, or other techniques.

GLBP (Gateway Load Balancing Protocol), a Cisco proprietary protocol that adds basic load balancing functionality to existing redundant router protocols.

Global Platform Device (GPD), a series of specifications for single and multi-application smart cards, acceptance devices and systems infrastructure. For more information see www.globalplatform.org.

Global Positioning System a system that uses signals from multiple satellites to determine locations in terms of longitude, latitude, and altitude.

Global Warming Potential (GWP), an estimate of how much a greenhouse gas contributes to global warming over a specific time interval. This is a relative scale that compares other gasses to the same mass of carbon dioxide, by definition the global warming potential of carbon dioxide is one. See Fig. 3 on page 280 for example values.

GLS Global Logistics System.

GMC (**General Motion Control**) technology used for manufacturing automation including general purpose motion controller systems and devices like servo and stepper drives or motors.

GMP (**Good Manufacturing Practice**) regulations and guidelines for the manufacture of drugs, medical devices, diagnostic products, and food that require documented procedures for production and testing, traceability of ingredients and the ability to recall product in the event of problems.

GMT Greenwich Mean Time, see **UTC**.

GNMP Government Network Management Profile, US standard FIPS 179-1.

GNOME a desktop software environment and development platform for **Linux** or **Unix** systems. See www.gnome.org for details.

GNU GPL (**General Public License**) or **GNU General Public License**, a widely-used **Open Source** software license developed for the GNU project. The GPL grants the rights of the free software definition and uses **Copyleft** to ensure these freedoms are preserved, even when the work is changed or extended; see www.gnu.org/licenses/ for more information.

Good Manufacturing Practice regulations and guidelines for the manufacture of drugs, medical devices, diagnostic products, and food that require documented procedures for production and testing, traceability of ingredients and the ability to recall product in the event of problems.

GOSIP Government Systems Interconnect Protocol, US standard requiring **OSI** protocol, replaced by **POSIT**

Governance business processes for developing and managing consistent, cohesive policies, processes and decision rights for a given area of responsibility. Corporate governance is the manner in which boards direct a corporation and the laws or customs applying to that direction.

Governance, Risk and Compliance business processes and systems for identifying and quantifying expected risks, mitigating these risks, and monitoring these activities. See also **Governance, Risk**, and **Compliance**.

GP Gauge Pressure.

GPD (Global Platform Device), a series of specifications for single and multi-application smart cards, acceptance devices and systems infrastructure. For more information see www.globalplatform.org.

GPL (General Public License) or **GNU General Public License**, a widely-used **Open Source** software license developed for the GNU project. The GPL grants the rights of the free software definition and uses **Copyleft** to ensure these freedoms are preserved, even when the work is changed or extended; see www.gnu.org/licenses/ for more information.

GPRS General Packet Radio Service, a mobile data service available on **GSM** or **TDMA/IS-136** cellular telephone networks.

GPS (Global Positioning System) a system that uses signals from multiple satellites to determine locations in terms of longitude, latitude, and altitude.

Graphics pictures and illustrations, visual representation or presentation of data, designs, etc. A graphic may be stored and coded as a **bitmap** or as a mathematical formula or vector. Graphics or 'custom graphics' may be used to depict a building or portions of a building in terms the building operator understands including diagrams of floor plans and equipment with overlays that show relevant data and control elements or animations to indicate status.

Graphics Interchange Format a bit-mapped image format with a limited color range that is widely used for simple images and animations.

Gray Water rain water or water that has been used once and can be reused for other purposes. For example, runoff and water from sinks may be reused for flushing toilets or in cooling towers.

GRC (Governance, Risk and Compliance) business processes and systems for identifying and quantifying expected risks, mitigating these risks, and monitoring these activities. See also **GOVERNANCE, RISK**, and **COMPLIANCE**.

Green Building a building that is more energy efficient, has a smaller environmental impact and is healthier for the occupants. See also **US GREEN BUILDING COUNCIL**.

Green Energy energy from environmentally friendly sources. Typically, this refers to renewable and non-polluting energy sources.

Green Label a **CARPET AND RUG INSTITUTE** testing and labeling program for that establishes limits on **TOTAL VOLATILE ORGANIC COMPOUNDS** and **4-PCH**.

Green Power electrical power from environmentally friendly sources. Typically, this refers to renewable and non-polluting sources.

Green Roof a roof made of environmentally friendly materials and/or incorporating environmental features such as a system for capturing rainwater, solar panels, or vegetation and soil over a waterproof membrane.

Green Seal a non-profit organization that identifies and promotes environmentally responsible products and services through standard setting, certification, and other activities. For more information see www.greenseal.org.

GreenFormat a **CONSTRUCTION SPECIFICATIONS INSTITUTE** standard format for reporting information about the sustainable attributes of construction products and online database of product information. For more information see www.csinet.org or www.greenformat.com.

Greenhouse Gas a gas that contributes to the greenhouse effect such as carbon dioxide, methane, nitrous oxide, and ozone.

Greenscapes US EPA program promoting cost-efficient and environmentally friendly landscaping, see SMART LANDSCAPING.

Grey Water see GRAY WATER.

GridWise a consortium of public and private stakeholders working to develop a new electrical power grid for the twenty-first century. See www.gridwise.org for more information.

Group Controller lighting control for a group of fixtures serving a specific area.

GS (Green Seal) a non-profit organization that identifies and promotes environmentally responsible products and services through standard setting, certification, and other activities. For more information see www.greenseal.org.

GSA US General Services Administration.

GSF Gross Square Foot or Footage.

GSM Global System for Mobile Communications, mobile telephone standard used in various forms in many parts of the world. See www.gsmworld.com for more information.

GSP (Gas Service Provider), a ENERGY SERVICE PROVIDER offering natural gas in a competitive market.

GTIN Global Trade Item Numbers, collective term for **UPC**, **EAN**, and JAN article numbers.

Guest Management business processes or systems for tracking and managing visitors to a facility. The scope and type of guest management functions vary by industry, for office buildings this is primarily visitor approval and tracking, in the hospitality industry this term may include customer preference tracking and other services.

GUI Graphical User Interface, a USER INTERFACE where user may point, click, drag, etc.

GWAC GridWise Architecture Council, a team of experts guiding that design of a future, intelligent, transactive, GRIDWISE energy system. See www.gridwiseac.org for more information.

GWP (Global Warming Potential), an estimate of how much a greenhouse gas contributes to global warming over a specific time interval. This is a relative scale that compares other gasses to the same mass of carbon dioxide, by definition the global warming potential of carbon dioxide is one. See Fig. 3 on page 280 for example values.

H

H.323 an **ITU** protocol recommendation for audio-visual session management. Used in **VOIP** and IP-based videoconferencing.

H/C Humidity Control.

H/W Hardware.

HA (High Availability) a system or device with features to enhance AVAILABILITY, usually through fault tolerance.

HAL (Hardware Abstraction Layer) an idealized software interface designed to hide the differences between different computer system implementations from the operating system and other software.

Halon a series of compounds consisting of bromine, fluorine, and carbon that were used as a fire extinguishing agents before they were banned because of their ozone depleting effects.

Hands-Free technology that can be used without touch. Examples include hands-free controls for sinks or bathroom fixtures.

Haptic pertaining to the sense of touch. A haptic user interface provides feedback through force, resistance, vibration or motion.

Hardscape non-plant elements of the landscape including patios, walkways, driveways, retaining walls, decks, water features, etc.

Hardware Abstraction Layer an idealized software interface designed to hide the differences between different computer system implementations from the operating system and other software.

Hardware Security Module (HSM), a computer device designed to generate, store, and physically protect keys and other secrets for use in cryptography. Many HSM devices are also hardware accelerators for cryptographic calculations.

HART (**Highway Addressable Remote Transducer**), communications technology for process instrumentation that allows two-way digital communication and analog signaling. For more information see www.hartcomm.org.

Hazard & Operability a qualitative method for examining processes to identify and evaluate potential problems, formalized in IEC 61882.

HAZOP (**Hazard & Operability**) a qualitative method for examining processes to identify and evaluate potential problems, formalized in IEC 61882.

HB Healthy Building.

HC or HCC, (**Horizontal Cross-Connect**) a STRUCTURED CABLING SYSTEM termination point (PATCH PANEL, punch down block, or CROSS-CONNECT) where vertical or backbone cabling of the CORE network connects to horizontal cabling of the EDGE NETWORK. Sometimes called a Floor Distributor.

HCFC (**Hydrochlorofluorocarbons**), chemical compounds consisting of hydrogen, chlorine, fluorine, and carbon that have been used as substitutes for CHLOROFLUOROCARBONS because of their reduced ozone depleting effects and are now being phased out in favor of Hydrofluorocarbons. See the HCFC classification in Fig. 3 on page 280.

HCI Human Computer Interface, another term for USER INTERFACE.

HCP (**Horizontal Connection Point**), an enhanced version of the CONSOLIDATION POINT defined in the **TIA-862** standard that may include a cross-connect and connections to other low-voltage building systems in addition to voice, data, and video.

HD-DVD High Density or High Definition DVD, an enhanced **DVD** format that can store 30 GIGABYTEs on a side (in two layers).

HDA (**Horizontal Distribution Area**), an area in a data center that serves as the distribution point for horizontal cabling and houses HORIZONTAL CROSS-CONNECTs and equipment for connecting to the ZONE DISTRIBUTION AREA or EQUIPMENT DISTRIBUTION AREA.

HDCP High-bandwidth Digital Content Protection, a technology developed by Digital Content Protection, LLC (a subsidiary of Intel) to protect digital entertainment content. HDCP has been implemented across both **DVI** and **HDMI** interfaces.

HDD Hard Disk Drive.

HDLC (**High Level Data Link Control**) ITU standard synchronous data link layer protocol; currently standardized as ISO 13239.

HDMI High-Definition Multimedia Interface, a single cable, all-digital audio/video interface standard for **HDTV**. HDMI supports standard, enhanced, or high-definition video (**DVI**), compressed or uncompressed multi-channel digital audio, and copy protection under the control of the video source.

HDTV (**High Definition TeleVision**) a digital television broadcasting system with higher resolutions (for example, **1080p**) than traditional formats (**NTSC, SECAM, PAL**). Compared to standard television, high-definition provides more detailed images (higher linear resolution) and less flicker (**Progressive Scanning**), additionally HDTV can display 16:9 aspect ratio pictures without **letterboxing**.

HDWE Hardware.

Heartbeat or **Heartbeat Message,** a periodic signal provided by a device and used to monitor operation and communication; problems are indicated by the absence of a heartbeat.

Heat Exchanger a **HVAC** component that transfers heat between physically separated fluids or airflows. In a chiller, the typical heat exchangers are the evaporator which exchanges heat between the **refrigerant** and the chilled water and the condenser which exchange heat between the refrigerant and condenser water.

Heat Recovery Ventilator a **HVAC** component that saves energy by using a **heat exchanger** to transfer heat from the warm building exhaust air to the cold outside air entering the building.

Heliostat a device that tracks the movement of the sun. Sun tracking can be used to maximize **Solar Panel** efficiency or drive **Daylighting** reflectors.

Help Desk an organization or system for technical support and customer service.

Hermetic Motor a motor that is sealed within the refrigerant atmosphere of a chiller and cooled by liquid **refrigerant**. A hermetic motor can be smaller and lighter than a comparable air-cooled motor.

HERS (**Home Energy Rating System**) California PUC program to certify home energy rating services. For more information see www.energy.ca.gov/HERS/.

HFC HYBRID FIBER COAX or **HYDROFLUOROCARBONS**.

HHD (Hybrid Hard Drive) or **Hybrid Hard Disk,** a disk drive equipped with a large buffer of non-volatile **FLASH** memory that is used to cache data during normal operation so that the mechanical disk can be turned off. The benefits of this design include faster start-up for the drive and the computer, decreased power requirements, and improved reliability.

HID (High-Intensity Discharge), a class of high output electrical light that includes mercury vapor, metal halide, sodium, and xenon short-arc lamps. In general these lamps are high efficiency, require a ballast, may have a warm-up period, and vary in their color characteristics.

Hierarchical Storage Management (HSM), a data storage technique which automatically moves data between high-cost fast access media and lower-cost and slower storage media.

High Availability a system or device with features to enhance **AVAIL-ABILITY,** usually through fault tolerance.

High-Definition Television a digital television broadcasting system with higher resolutions (for example, **1080P**) than traditional formats (**NTSC, SECAM, PAL**). Compared to standard television, high-definition provides more detailed images (higher linear resolution) and less flicker (**PROGRESSIVE SCANNING**), additionally HDTV can display 16:9 aspect ratio pictures without **LETTERBOXING**.

High-Intensity Discharge (HID), a class of high output electrical light that includes mercury vapor, metal halide, sodium, and xenon short-arc lamps. In general these lamps are high efficiency, require a ballast, may have a warm-up period, and vary in their color characteristics.

High Level Data Link Control ITU standard synchronous data link layer protocol; currently standardized as ISO 13239.

High-Performance Façade building exterior designed to support daylighting, solar heat gain control, ventilation and space conditioning.

High Power Factor Ballast a BALLAST a power factor greater than 0.9.

High Speed Downlink Packet Access (HSDPA), a **HSPA** mobile telephony protocol that provides DOWNLINK data transfer speeds as high as 14.4 Mbit/s using GSM cell phone networks. For more information see www.gsmworld.com under 3GSM.

High Speed Packet Access (HSPA), a family of mobile telephony protocols designed to improve the performance of **UMTS** networks; initial protocols include **HSDPA, HSUPA**, and **HSPA EVOLVED**. For more information see hspa.gsmworld.com.

High Speed Uplink Packet Access a **HSPA** mobile telephony protocol designed to provide UPLINK speeds up of 5.76 Mbit/s or higher.

High Voltage commonly used to mean 110 volts or higher, contrast to LOW VOLTAGE; the NATIONAL ELECTRICAL CODE defines high voltage as over 600 volts.

Highway Addressable Remote Transducer (HART), communications technology for process instrumentation that allows two-way digital communication and analog signaling. For more information see www.hartcomm.org.

HIS Hospital Information System.

HL Horizontal Link.

HMI Human Machine Interface, see USER INTERFACE.

Hockey Puck slang for a location tracking transmitter device.

HOE (Holographic Optical Element), a hologram that acts as a lens or mirror with the ability to focus, diffuse, or redirect light.

Holographic Optical Element (HOE), a hologram that acts as a lens or mirror with the ability to focus, diffuse, or redirect light.

Home Energy Rating System California PUC program to certify home energy rating services. For more information see www.energy.ca.gov/HERS/.

Home Run Cabling cabling with a dedicated cable going from the TELECOMMUNICATIONS ROOM to each work area. Contrast to ZONE CABLING.

Homeland Security Presidential Directive12 Policy for a Common Identification Standard for Federal Employees and Contractors mandate for standard credentials (see **FIPS 201**) and changes to the credential process.

Horizontal Connection Point (HCP), an enhanced version of the CONSOLIDATION POINT defined in the TIA-862 standard that may include a cross-connect and connections to other low-voltage building systems in addition to voice, data, and video.

Horizontal Cross-Connect a STRUCTURED CABLING SYSTEM termination point (PATCH PANEL, punch down block, or CROSS-CONNECT) where vertical or backbone cabling of the CORE network connects to horizontal cabling of the EDGE NETWORK. Sometimes called a Floor Distributor.

Horizontal Distribution Area (HDA), an area in a data center that serves as the distribution point for horizontal cabling and houses HORIZONTAL CROSS-CONNECTs and equipment for connecting to the ZONE DISTRIBUTION AREA or EQUIPMENT DISTRIBUTION AREA.

Host a computer that can be accessed over a network, a SERVER. A Web hosting company provides servers, Internet connectivity and other services.

Hot and Cold Aisles a data center layout technique designed to increase equipment cooling effectiveness by arranging cabinets and racks in an alternating pattern with fronts facing for a cool aisle and backs facing for a hot aisle. This minimizes the extent to which the hot exhaust air from one cabinet becomes the intake air for other cabinets.

Hot Standby Router Protocol (HSRP), a Cisco protocol for establishing a fault-tolerant default gateway as part of a redundant network, standardized as RFC 2281. See also VIRTUAL ROUTER REDUNDANCY PROTOCOL.

Hotel Technology Next Generation a trade association facilitating development of next generation technologies for the hospitality industry. See www.htng.org for more information.

Hoteling or **Hot Desk**, flexible on-demand workspace assignment schemes for office workers designed to reduce total space requirements and costs. Workers who are not in the office full time share a pool of workstations on an as-needed basis.

HOV High-Occupancy Vehicle.

HP Horsepower.

HPF Ballast (High Power Factor Ballast) a BALLAST with a power factor greater than 0.9.

HPI Hydrocarbon Processing Industry.

HPM Hazardous Production Materials.

HR (Human Resources) business processes and systems for tracking and managing employees and other personnel. Functions may include time and attendance, payroll, benefits administration, skills inventory, employee development, etc.

Href Hypertext reference, the target specified in a hypertext link.

HRV (Heat Recovery Ventilator) a HVAC component that saves energy by using a HEAT EXCHANGER to transfer heat from the warm building exhaust air to the cold outside air entering the building.

HSDPA (High Speed Downlink Packet Access), a **HSPA** mobile telephony protocol that provides DOWNLINK data transfer speeds as high as 14.4 Mbit/s using GSM cell phone networks. For more information see www.gsmworld.com under 3GSM.

HSM HARDWARE SECURITY MODULE or **HIERARCHICAL STORAGE MANAGEMENT.**

HSPA (High Speed Packet Access), a family of mobile telephony protocols designed to improve the performance of **UMTS** networks; initial protocols include **HSDPA, HSUPA,** and **HSPA EVOLVED.** For more information see hspa.gsmworld.com.

HSPA Evolved an upgrade to **HSPA** networks designed to provide higher data rates (42Mbit/s or more in DOWNLINK) and greater network capacity.

HSRP (Hot Standby Router Protocol), a Cisco protocol for establishing a fault-tolerant default gateway as part of a redundant network, standardized as RFC 2281. See also **VIRTUAL ROUTER REDUNDANCY PROTOCOL.**

HSUPA (High Speed Uplink Packet Access) a **HSPA** mobile telephony protocol designed to provide UPLINK speeds up of 5.76 Mbit/s or higher.

Htg Heating.

HtgSP Heating SETPOINT.

HTIB Home Theater In-a-Box, a prepackaged home theater system that typically contains everything but the display/TV.

HTML HyperText Markup Language, the standard coding used to identify the elements of a Web page and control some aspects of display formatting.

HTNG (Hotel Technology Next Generation) a trade association facilitating development of next generation technologies for the hospitality industry. See www.htng.org for more information.

HTTP HyperText Transport Protocol, the request/response protocol used to access Internet Web pages.

HTTPS is the encrypted version of **HTTP** using **SSL** or **TLS**.

Human Resources business processes and systems for tracking and managing employees and other personnel. Functions may include time and attendance, payroll, benefits administration, skills inventory, employee development, etc.

Hunt Group a group of associated telephone numbers where an incoming call is automatically routed to an idle (not busy) telephone line for completion.

HVAC Heating, Ventilating, and Air Conditioning.

HVAC/R Heating, Ventilating, Air Conditioning and Refrigeration.

HW Hot Water.

Hybrid Fiber Coax or **Hybrid Fibre-Coaxial,** any network design that includes a mix of high speed fiber optic networking as the backbone and less expensive coaxial cable at the edges of the network. This type of design has been used for CABLE TV and CCTV networks.

Hybrid Hard Drive or **Hybrid Hard Disk,** a disk drive equipped with a large buffer of non-volatile **FLASH** memory that is used to cache data during normal operation so that the mechanical disk can be turned off. The benefits of this design include faster start-up for the drive and the computer, decreased power requirements, and improved reliability.

Hydrochlorofluorocarbons (HCFC), chemical compounds consisting of hydrogen, chlorine, fluorine, and carbon that have been used as substitutes for CHLOROFLUOROCARBONS because of their reduced ozone depleting effects and are now being phased out in favor of Hydrofluorocarbons. See the HCFC classification in Fig. 3 on page 280.

Hydrofluorocarbons (HFC), compounds consisting of hydrogen, fluorine, and carbon that are being used as replacements for **CHLOROFLUORO-CARBONS** because they do not contain chlorine or bromine and do not deplete the ozone layer. See the HFC classification in Fig. 3 on page 280.

Hydronics the use of water as the heat-transfer medium for heating and cooling systems. See **CHILLED BEAMS, DISTRICT HEATING AND/OR COOLING, UNDERFLOOR HEATING,** etc.

Hygrometer an instrument used to measure the amount of moisture in the air or humidity.

Hyperlink a navigation element in a document that may reference another section of the same document, another document, or a specified section of another document, and will automatically display the referred information to the user when the navigation element is selected. Hyperlinks are a foundation of the **WEB** but may appear in any electronic media.

Hypertext documents that contain automatic cross-references to other documents called hyperlinks. Selecting a **HYPERLINK** causes linked document to load and display.

I

I&A (Identification and Authentication) a two-step **ACCESS CONTROL** process. Identification should uniquely identify a user or a group of users; common examples include cards, account numbers, user names, computer **MAC ADDRESS,** etc. Authentication is a process for verifying the claimed identity based on something you know, such as a password or a personal identification number; something you have such as a **SMART CARD,** token, or key; or something you are, such your picture or a **BIOMETRIC** such as fingerprint, voice, retina, or iris characteristics.

I/O Input/Output.

IA Industrial Automation or Inspection Agency.

IAI (International Alliance for Interoperability) a group of organizations working together to improve the productivity and efficiency of the construction and facilities management industries through **BUILDING INFORMATION MODEL** standards. See www.iai-international.org and www.iai-na.org for more information.

IANA Internet Assigned Numbers Authority, an agency operated by **ICANN** that oversees global IP address allocation, **DNS** root zone management, and other Internet protocol assignments.

IAQ (Indoor Air Quality), the healthfulness of air inside a building; indoor air quality issues include temperature, humidity, and carbon dioxide levels as well as pollution, mold, smoke, radon, etc. The term IAQ sensor is sometimes used for carbon dioxide sensors, sensors that measure **VOLATILE ORGANIC COMPOUNDS**, or the combination of carbon dioxide, temperature and humidity.

IB (InfiniBand) a very high speed interconnection technology (2-96 Gbit/s) used to connecting together the components of high-performance computing clusters, supercomputer interconnects and for inter-switch connections.

IBIIS Internet Based Integrated Information Systems.

IBS Intelligent Building System or Systems, general term for the systems in an **INTELLIGENT BUILDING**.

IBW In Building Wireless, see **DISTRIBUTED ANTENNA SYSTEM**.

IC Integrated Circuit or **INTERMEDIATE CROSS-CONNECT**.

iCalendar or **RFC 2445**, a standard for calendar data exchange via email (as a **MIME** content type) that supports scheduling events, setting alarms, assigning to-do or action items, making journal entries, and publishing or exchanging free/busy information.

ICANN Internet Corporation for Assigned Names and Numbers, the organization that manages centrally coordinated identifiers for the Internet under contract with the US Department of Commerce. For more information see www.icann.org.

ICC Intermediate Cross-Connect or International Code Council.

ICIA see InfoComm.

ICRA Infection-Control Risk Assessment or Internet Content Rating Association.

ICT Information and Communications Technology.

iDEN (Integrated Digital Enhanced Network) a mobile telecommunications technology developed by Motorola and used for public and private networks.

Identification and Authentication a two-step access control process. Identification should uniquely identify a user or a group of users; common examples include cards, account numbers, user names, computer MAC address, etc. Authentication is a process for verifying the claimed identity based on something you know, such as a password or a personal identification number; something you have such as a Smart Card, token, or key; or something you are, such your picture or a biometric such as fingerprint, voice, retina, or iris characteristics.

Identity Management System a computer application for enrolling and managing users as part of a Smart Card or Personal Identity Verification scheme.

IDF Intermediate Distribution Frame, telephone central office equivalent of ICC.

IDMS (Identity Management System) a computer application for enrolling and managing users as part of a Smart Card or Personal Identity Verification scheme.

IDS (Intrusion Detection System) 1) a *physical security* system for detecting unauthorized entry or access; 2) a *network or information security system* designed to detect actions or evidence of attempts to compromise the confidentiality, integrity or availability of a resource.

IEA or **International Energy Agency,** a non-governmental energy research agency and energy policy advisor with 26 Member countries. For more information see www.iea.org.

IEC International Engineering Consortium or International Electrotechnical Commission, international standards body for electrotechnology.

IED Improvised Explosive Device.

IEEE Institute of Electrical and Electronics Engineers, engineering professional organization that, among other things, develops radio and communications protocol standards and submits them to ANSI for adoption as American National Standards and international standards.

IEEE 802.5 or IEEE 802.5, standard for TOKEN RING LOCAL AREA NETWORK.

IEQ (Indoor Environmental Quality) the total indoor experience including thermal comfort, INDOOR AIR QUALITY (IAQ), lighting, acoustics, etc.

IESNA Illuminating Engineering Society of North America, a lighting industry organization involved in lighting standards development and education for electrical lighting and DAYLIGHTING. See www.iesna.org for more information.

IETF (Internet Engineering Task Force) the group that establishes technical standards for the Internet including the TCP/IP protocol suite. See www.ietf.org for more information.

IFC (Industry Foundation Classes) an object based data model for the AEC industry published by the IAI.

ifcXML the XML based implementation of the **IFC** object model.

IFMA see INTERNATIONAL FACILITY MANAGEMENT ASSOCIATION.

IFP Intelligent Front Panel.

IG Isolated Ground.

IGES (**Initial Graphics Exchange Specification**) a neutral data format for transferring **CAD** models. See www.nist.gov/iges/ for more information.

IIB Intelligent and Integrated Building, another term for INTELLIGENT BUILDING.

IIS (**Internet Information Services**) (formerly Internet Information Server), Microsoft Web Server software that runs on Microsoft Windows.

ILM (**Infrastructure Lifecycle Management**) processes and software applications to support the management of buildings over their useful life including design, construction, and operation from a single system of record.

IM Instant Message or INSTANT MESSAGING, online communication using brief messages exchanged in real-time.

Image Map 1) a graphic or image with one or more hyperlinks or interactive areas; 2) the data structure that relates areas of an image to hyperlink target addresses or actions.

IMAP Internet Message Access Protocol, an application level protocol for accessing e-mail messages.

Independent Representative a sales representative that is not an employee.

Indoor Air Quality (IAQ), the healthfulness of air inside a building; indoor air quality issues include temperature, humidity, and carbon dioxide levels as well as pollution, mold, smoke, radon, etc. The term

IAQ sensor is sometimes used for carbon dioxide sensors, sensors that measure VOLATILE ORGANIC COMPOUNDS, or the combination of carbon dioxide, temperature and humidity.

Indoor Environmental Quality the total indoor experience including thermal comfort, INDOOR AIR QUALITY (IAQ), lighting, acoustics, etc.

Industrial Ethernet standards and technology for Ethernet in industrial applications with rugged hardware and high reliability features. See www.odva.org for more information.

Industry Foundation Classes an object based data model for the **AEC** industry published by the **IAI**.

Infant Abduction System an ACTIVE **RFID** tracking application for medical facilities designed to prevent the unauthorized removal of infants by tracking the infant using a wrist or ankle band and automatically locking doors or alarming if the infant moves outside of a defined area without being accompanied by an authorized person.

Infection Control use of technology to combat infectious disease and reduce the number of healthcare-associated infections. For example, ventilation systems designed to minimize the spread of airborne germs.

Infection-Control Risk Assessment (ICRA), a multidisciplinary process for reducing risk from infection throughout facility planning, design, and construction activities. JOINT COMMISSION standards require hospitals to maintain an Infection Control Risk Assessment program for all construction; the **AIA** publishes guidelines for this process.

Infectious Disease a disease caused by organisms that can multiply within a body. Contrast to CONTAGIOUS DISEASE; a disease may be infectious but not contagious.

InfiniBand a very high speed interconnection technology (2-96 Gbit/s) used to connecting together the components of high-performance computing clusters, supercomputer interconnects and for inter-switch connections.

InfoComm the International Communications Industry Association, an audiovisual trade association; see www.infocomm.org for more information.

Information Systems Audit and Control Association (ISACA), membership and research organization promoting IT governance and control that helped develop and publishes **COBIT**. For more information see www.isaca.org.

Information Technology Infrastructure Library (ITIL), a cohesive set of best practice documents for IT service management and an associated professional credentialing scheme. See www.itil.co.uk for details.

InfraRed electromagnetic radiation with wavelengths longer than visible light but shorter than microwave, used for communications and specialized imaging. See also ACTIVE IR and PASSIVE IR.

Infrastructure Lifecycle Management processes and software applications to support the management of buildings over their useful life including design, construction, and operation from a single system of record.

Inheritance in object-oriented programming (see **OO**) inheritance provides a mechanism for creating a subclass that is an extension of a more general class of objects while sharing (inheriting) some properties and behaviors of the parent class.

Initial Graphics Exchange Specification a neutral data format for transferring **CAD** models. See www.nist.gov/iges/ for more information.

Initiating Device the system component that originates transmission of a change-of-state condition, or alarm signal, such as a smoke detector, manual fire alarm box, or supervisory switch.

Instant Messaging a form of near real-time communication between two or more people based on typed text, may also include audio, video, file transfer and other capabilities.

Institute for Research in Construction of the Canadian National Research Council, see www.irc.nrc-cnrc.gc.ca for more information.

Institute of Management Administration a business management research and networking organization. For more information see www.ioma.com.

Integrated Digital Enhanced Network a mobile telecommunications technology developed by Motorola and used for public and private networks.

Integrated Façade a FAÇADE that is designed, analyzed and procured as a system taking designed to support daylighting, solar heat gain control, ventilation and space conditioning.

Integrated Practice a term used by the AIA and others to describe new architecture service delivery models that encompass the entire project lifecycle and are enabled by BUILDING INFORMATION MODELS and other new technologies. For more information see www.aia.org/ip_default.

Integrated Workplace Management Systems a computer application for all workplace related assets and functions including facilities management, corporate real estate management, project management, and operations.

Integration the process of combining or accumulating. Integrating systems brings together the data and services of multiple applications and systems to enable new functionality and greater productivity for occupants and owner/operators.

Intelligent Bathroom a bathroom equipped with interconnected devices to control lighting, temperature, humidity, etc. May also include remote monitoring of soap dispensers, paper towel and toilet paper dispensers, trash receptacles, and water flow sensors, multimedia information and entertainment, or biosensors for occupant health monitoring. See also INTELLIGENT TOILET.

Intelligent Building a building that integrates technology and process to create a facility that is safer, more comfortable and productive for its occupants, and more operationally efficient for its owners. Advanced technology—combined with improved processes for design, construction and operations—provide a superior indoor environment that improves occupant comfort and productivity while reducing energy consumption and operations staffing..

Intelligent Toilet a toilet equipped with sensors that enable it to measure and monitor the user's weight, body fat, urine sugar levels, etc.

Interactive Program Guide a dynamic guide to available content or channels that allows a viewer to find specific content based on various criteria. For example, a Cable TV program guide typically displays shows by genre or channel and time.

Interactive Voice Response systems that enable users to interact with computer systems over the telephone. Features may include input based on telephone network signaling (including the calling number or Automatic Number Identification), keypad (DTMF input), and speech recognition; and output from pre-recorded messages, synthesized speech, Text-To-Speech, etc.

Interconnect a cable with pre-attached plugs or connectors for connecting compatible audio or video equipment.

Interlace or **Interlacing,** a technique for displaying video images that requires two cycles to change the whole display by refreshing the even-numbered lines and then the odd-numbered lines.

Intermediate Cross-Connect or **Building Distributor,** an optional connection point between the Main Cross-Connect (first-level backbone) and the Horizontal Cross-Connect (second-level backbone) as part of a Structured Cabling System.

International Alliance for Interoperability a group of organizations working together to improve the productivity and efficiency of the construction and facilities management industries through Building Information Model standards. See www.iai-international.org and www.iai-na.org for more information.

International Code Council (ICC), a membership organization dedicated to building safety and fire prevention that develops a range of International Codes. ICC was founded by Building Officials and Code Administrators International (BOCA), International Conference of Building Officials (ICBO), and Southern Building Code Congress International (SBCCI). For more information see www.iccsafe.org.

International Commission on Illumination or **Commission Internationale de l'Eclairage** (CIE), professional and standards developing organization for the science and art of lighting. See www.cie.co.at for more information.

International Facility Management Association (IFMA), a professional organization for facilities managers. The IFMA certifies facility managers, conducts research, provides educational programs, recognizes degree and certificate programs and produces facility management-related conferences, see www.ifma.org for more information.

International Performance Measurement and Verification Protocol provides an overview of current best practice techniques available for measuring and verifying the results of energy efficiency, water efficiency, and renewable energy projects in commercial and industrial facilities. For more information see www.evo-world.org.

International Telecommunication Union a specialized agency of the United Nations that publishes international standards.

Internet 1) the global public TCP/IP network of loosely connected and diverse computers, the **WORLD WIDE WEB** and e-mail are only a few of the applications available on this network; 2) any public or private subnetwork based on **INTERNET PROTOCOLS**.

Internet Engineering Task Force the group that establishes technical standards for the Internet including the TCP/IP protocol suite. See www.ietf.org for more information.

Internet Information Services (formerly Internet Information Server), Microsoft Web Server software that runs on Microsoft Windows.

Internet Protocol the low level (network layer) protocol in the Internet (or TCP/IP) protocol suite, most commonly IPv4 or RFC 791. In addition to providing for packet delivery, the IP protocol also provides a global addressing service with 32 bit addresses in **IPv4** and 128 bit addresses in **IPv6**.

Internet Protocols or **Internet Protocol Stack**, a general term for the suite of **TCP/IP** based protocols formalized by the **IETF** and utilized by the **INTERNET**. See Fig. 7 on page 283.

Internet Relay Chat a protocol for real-time chat or synchronous conferencing designed for group discussion in forums called channels; also allows one-to-one communication and data transfers via private message.

Internet TV a general term for television service delivered over an **IP** network which may or may not be the public **INTERNET** and may include access to a variety of different functions and sources of content. See also **IPTV**.

Intranet an internal network that uses **INTERNET PROTOCOLS** and technologies.

Intrusion Detection System 1) a *physical security* system for detecting unauthorized entry or access; 2) a *network or information security system* designed to detect actions or evidence of attempts to compromise the confidentiality, integrity or availability of a resource.

Intumescent Paint a type of fire suppressant paint coating.

iODBC an **OPEN SOURCE** and platform-independent software library for **ODBC**. For more information see www.iodbc.org.

IOMA (Institute of Management Administration) a business management research and networking organization. For more information see www.ioma.com.

IOUM Innovation in Operations, Upgrades and Maintenance.

IP Ingress Protection see **IP Code, Integrated Practice** or **Internet Protocol**.

IP Address a unique numeric identifier assigned to each device in an **Internet Protocol** network. Addresses may be fixed (static) or dynamic, see also **DHCP**.

IP CCTV Closed-Circuit Television connected via shared IP networks.

IP Code or **IP Rating,** Ingress Protection Ratings defined in IEC 60529 classify the level of protection that an electrical appliances provide against the intrusion of solid objects or dust, accidental contact, and water. Similar to NEMA Ratings.

IP Enabled a device that has the ability to communicate using Internet Protocol or over an **IP** network.

IP Geolocation a process for determining the approximate geographic location of an Internet connected computer based on its Internet Protocol address.

IP Media Internet Protocol Media, broadband distribution of audio, video, and multimedia content.

IPBX Intranet Private Branch eXchange, a **private branch exchange** or switching system for **VOIP**.

IPC Interprocessor Communication between processors within system or cluster or Inter Process Communications, between collaborating processes within a system.

IPG Interactive Program Guide or **IP Geolocation.**

IPICS IP Interoperability and Collaboration System, a Cisco technology for bridging multiple types of radios (including **LMR**), telephones, mobile phones, **VOIP** phones, intercoms, etc. over a shared **IP** network.

IPM Integrated Pest Management, programs that minimize the use of pesticides by reducing sources of food, water and shelter for pests.

IPMI Intelligent Platform Management Interface, specifications for common monitoring interfaces to computer server hardware and firmware that can be used to monitor system health and manage the system.

IPMVP (International Performance Measurement and Verification Protocol) provides an overview of current best practice techniques available for measuring and verifying the results of energy efficiency, water efficiency, and renewable energy projects in commercial and industrial facilities. For more information see www.evo-world.org.

IPO Initial Public Offering, making the stock of a company available for sale to the general public.

IPR Intellectual Property Rights.

IPS Internet Protocol Suite, **TCP/IP** and related protocols.

IPsec IP security, a network layer protocol standard for securing IP communications by encrypting and/or authenticating IP packets. For more information see RFC 4301–4309.

IPTV Internet Protocol Television, television transmitted using IP networks.

IPv4 Internet Protocol version 4, the most commonly used version of **INTERNET PROTOCOL**, the network layer protocol in the **INTERNET PROTOCOLS** stack for both the public **INTERNET** and private networks. IPv4 uses 32-bit addresses and is being replaced by **IPv6** which supports 128-bit addresses in the backbone of the Internet and very large private networks.

IPv6 IP version 6, an update to the **IP** protocol specification that supports larger networks and other features via expanded IP address fields.

IR INDEPENDENT REPRESENTATIVE or **INFRARED**.

IRC Institute for Research in Construction or Internet Relay Chat.

IrDA Infrared Data Association, a trade organization that defines physical specifications and protocol standards for the exchange of data over infrared.

IS Information System.

ISA the Instrumentation, Systems, and Automation Society, formerly the Instrumentation Society of America. See www.isa.org for more information.

ISACA (Information Systems Audit and Control Association), membership and research organization promoting IT governance and control that helped develop and publishes **COBIT**. For more information see www.isaca.org.

ISC2 or **ISC²**, International Information Systems Security Certification Consortium, a nonprofit organization that maintains industry best practices for information security and certifies information security professionals. For more information see www.isc2.org.

iSCSI standard for using the **SCSI** protocol over TCP/IP storage networks.

ISDN Integrated Services Digital Network, a CIRCUIT-SWITCHING telephone service that provides digital transmission of voice and data over ordinary telephone lines with better quality, higher speeds, and advanced call features.

ISFET Ion Sensitive Field Effect Transistor, a type of sensor used to measure ion concentration or pH in solution.

ISM Band or **ISM Frequencies**, Industrial, Scientific and Medical Band, a group of radio frequencies within several different ITU bands (HF, UHF, VHF, EHF, see Fig. 8 on page 284) that were originally reserved for purposes other than communications (including microwave ovens).

Some of these bands are now shared with communications equipment based on the understanding that communications gear must accept any interference generated by other ISM equipment.

ISO International Organization for Standardization, a non-governmental organization that publishes international standards, a federation of national standards bodies from more than 145 countries. See www.iso.ch for more information.

ISO/IEC-11801 Generic Customer Premises Cabling Standard, an international cabling standard based on ANSI/TIA/EIA-568, see **TIA-568**.

ISO/IEC 20000 an international standard for IT Service Management that incorporates the **INFORMATION TECHNOLOGY INFRASTRUCTURE LIBRARY** framework.

ISO 10303 see **STANDARD FOR THE EXCHANGE OF PRODUCT MODEL DATA**.

ISO 14000 a series of standards for environmental management intended to help organizations minimize the negative environmental impact of their operations, comply with applicable laws, regulations, and other requirements, and continually improve in the above. See www.iso.ch for more information.

ISO 9000 family of standards for quality management systems in manufacturing or service. These standards assure that consistent business processes are being applied, see www.iso.ch for more information.

ISP Internet Service Provider.

ISSA Information Systems Security Association, an international organization of information security professionals and practitioners. For more information see www.issa.org.

ISSAF Information Systems Security Assessment Framework, a structured framework for information security evaluation developed by the Open Information System Security Group.

ISV Independent Software Vendor.

IT Information Technology.

IT Centric centered on the needs, technical skills, or viewpoint of **IT** professionals.

IT Governance Institute (ITGI), a research think tank focused on IT-enabled business systems governance that helped develop **COBIT**. For more information see www.itgi.org.

ITGI (IT Governance Institute), a research think tank focused on IT-enabled business systems governance that helped develop **COBIT**. For more information see www.itgi.org.

ITIL (Information Technology Infrastructure Library), a cohesive set of best practice documents for IT service management and an associated professional credentialing scheme. See www.itil.co.uk for details.

ITS Information Transport Systems, cabling and wireless network infrastructure for communications, life safety, and automation.

ITT or Invitation To Tender, see **RFP**.

ITU (International Telecommunication Union) a specialized agency of the United Nations that publishes international standards.

IVR (Interactive Voice Response) systems that enable users to interact with computer systems over the telephone. Features may include input based on telephone network signaling (including the calling number or **AUTOMATIC NUMBER IDENTIFICATION**), keypad (**DTMF** input), and speech recognition; and output from pre-recorded messages, synthesized speech, **TEXT-TO-SPEECH**, etc.

IWMS (Integrated Workplace Management Systems) a computer application for all workplace related assets and functions including facilities management, corporate real estate management, project management, and operations.

J

J2ME or **Java ME,** Java 2 Platform, Micro Edition, a collection of **Java** software **APIs** to support the development of applications for resource-constrained devices such as PDAs, cell phones, games, set-top boxes, and other consumer appliances. Sun Microsystems has made the J2ME source code **Open Source** under the **General Public License** as part of the **phone**ME project.

Jace 1) Java Application Control Engine, Tridium Niagara area controller product line; 2) a set of open source software libraries that make it easier to execute C++ code within the **Java Virtual Machine** environment.

Java an object-oriented programming language for Internet applications.

Java Message Service a standard for distributed Java applications that are loosely coupled using a reliable asynchronous messaging infrastructure. See also **MQ.**

Java Virtual Machine a runtime environment or virtual machine that interprets and executes Java **bytecode.** Typically bytecode is generated by Java language compilers; it can be generated by compilers for other languages.

JavaScript ECMAScript, or **JScript** (Microsoft) are prototype-based programming language mostly used for client-side applications. Although JavaScript and Java both use C syntax they are very different languages.

JCAF Java Control & Automation Framework.

JCAHO or **Joint Commission on Accreditation of Healthcare Organizations,** see **Joint Commission.**

JDBC Java Database Connectivity, a JAVA language **API** for access to contents of a RELATIONAL DATABASE MANAGEMENT SYSTEM that provides a consistent interface to a variety of different database implementations including **ODBC** compliant databases.

JEIDA Japan Electric Industry Development Association.

JEIF Japan Electrical Industrial Federation.

JEMA Japan Electrical Manufacturers' Association

JEMIMA Japan Electrical Measuring Instruments Manufacturers' Association.

JISC Japanese Industrial Standards Committee.

JIT Just-In-Time.

Jitter variations in the timing of a signal or the latency of transmission.

JMS (Java Message Service) a standard for distributed Java applications that are loosely coupled using a reliable asynchronous messaging infrastructure. See also **MQ.**

Joint Commission formerly the **Joint Commission on Accreditation of Healthcare Organizations,** a nonprofit organization that evaluates and accredits health care organizations and programs in the United States. See www.jointcommission.org for more information.

JPEG Joint Photographic Experts Group, a standards group for image formats (see www.jpeg.org for more information). The term JPEG or JPG is more commonly used as shorthand for the compressed image formats defined by this group and formalized as ISO/IEC IS 10918-1.

Jumper see PATCH CABLE.

JVM (Java Virtual Machine) a runtime environment or virtual machine that interprets and executes Java BYTECODE. Typically bytecode is generated by Java language compilers; it can be generated by compilers for other languages.

K

KDE K Desktop Environment, a desktop software environment and development platform for LINUX, UNIX, and other systems.

Kerberos a computer network authentication protocol and supporting software published by Massachusetts Institute of Technology. See web.mit.edu/Kerberos for more information.

Kernel a central component of a computer OPERATING SYSTEM responsible for managing system resources and the interactions between hardware and software components.

Key or **Cryptographic Key**, a piece of information that controls the operation of a cryptographic ALGORITHM. The key specifies the particular transformation of plaintext into ciphertext during encryption, and vice versa during decryption.

Key Management System (KMS), the business processes and computer applications to generate, exchange, store, safeguard, vet, and replace Cryptographic KEYs. Key management that concerns keys at the user level may be separate from key scheduling and handling key material within the operation of a cipher system.

Key Recovery any cryptographic code breaking activity based on determining the key value from the encoded messages. A RELATED KEY ATTACK is one type of key recovery threat.

Keyword 1) a defined term in a computer language or command set; 2) a specific term related to a subject area or useful for retrieving information.

kHz Kilo Hertz, thousands of cycles per second.

Kiosk a standalone interactive display system, may be equipped with a TOUCH SCREEN, keyboard, video camera, microphone, etc.

KM (**Knowledge Management**) business processes and systems intended to identify, capture, organize, and distribute knowledge for reuse, awareness, and learning within an organization.

KML or **Keyhole Markup Language,** an **XML** based file format used to define three-dimensional geospatial data for display in an earth browser such as Google Earth or Google Maps. KML can specify PLACEMARKs, features (images, polygons, 3D models, textual descriptions, etc.), and camera views.

KMS KEY MANAGEMENT SYSTEM or KNOWLEDGE MANAGEMENT SYSTEM.

KMZ a compressed file format used to distribute **KML** data. Each KMZ file is a ZIP archive that contains a single KML file ("doc.kml") along with any overlay and icon images referenced in the KML.

Knowledge Management business processes and systems intended to identify, capture, organize, and distribute knowledge for reuse, awareness, and learning within an organization.

Knowledge Management System (KMS), business processes and computer applications for capturing and managing knowledge within an organization. Typically this includes support for knowledge creation, capture, storage and dissemination of expertise and knowledge.

KNX an OSI-based network communications protocol for intelligent buildings standardized as EN 50090 and ISO/IEC 14543. KNX is the successor to the European Home Systems Protocol (EHS), BatiBUS, and the European Installation Bus (EIB); see www.konnex.org for more information.

KPI Key Performance Indicator.

KVA Kilo Volt Ampere, power of one thousand volt amperes or one thousand watts with a purely resistive load. With more complex loads the apparent power used in volt amperes will be larger than the true power used in watts, the ratio of these two quantities is called the POWER FACTOR.

KVM Keyboard Video and Monitor, switches or systems that allow these devices to be shared between multiple servers or other computers.

KW Kilowatt, power of one thousand watts.

L

LAN (Local Area Network) a network within a building or campus. Contrast to METROPOLITAN AREA NETWORK or WIDE AREA NETWORK.

Language a system used for communication with a finite set of symbols and a defined set of rules or grammar. May be used to refer to a programming language, markup language, interface messages or a PROTOCOL.

LCA (Life Cycle Assessment) a process for evaluating the environmental impact of building materials over the entire life of the building including disposal.

LCC (Life Cycle Cost) total economic cost of a resource, may include cost to create/build, own/operate, and destruction/disposal costs. See also LIFE CYCLE ASSESSMENT.

LCCC (Life Cycle Cost Calculator) financial tool for calculating the economic cost of ownership of a resource or building, see LIFE CYCLE COST.

LCD (Liquid Crystal Display) a display technology that uses transmitted or reflected light, polarizing filters, and a type of semiconductor material that can be electrically switched to allow or block light transmission; also used for electrically controlled privacy glass. For color displays each PIXEL contains three cells (subpixels) with red, green, and blue filters.

LCI Load Commutated Inverter, an inverter that converts direct current to alternating current at a range of different frequencies.

LCoS (Liquid Crystal on Silicon) a type of LIQUID CRYSTAL DISPLAY built on a silicon surface with a reflective coating of aluminum. This technology allows smaller pixels and better contrast than transmissive liquid crystal displays.

LDAP Lightweight Directory Access Protocol, a protocol for querying and updating directory services over TCP/IP networks. LDAP directories follow the **X.500** model.

Leadership in Energy and Environmental Design a rating system developed by the US Green Building Council. See www.usgbc.org/LEED/ for more information.

LEAP or **Lightweight Extensible Authentication Protocol,** a proprietary WIRELESS LOCAL AREA NETWORK authentication method developed by Cisco Systems that uses dynamic **WEP** keys and mutual authentication between a wireless client and a **RADIUS** server.

LED (Light-Emitting Diode) semiconductor technology used for indicators, lighting, etc.

LEED (Leadership in Energy and Environmental Design) a rating system developed by the US Green Building Council. See www.usgbc.org/LEED/ for more information.

LEED AP LEED Accredited Professional, certification for design, construction, operation and maintenance professionals based on green building practices and principles and familiarity with LEED requirements, resources, and processes.

Legacy System an existing computer system or application that continues to be used even though it may be outdated.

LES Logistics Execution System, a supply chain management application that coordinates production, procurement, storage, distribution, and transportation.

Letterboxing the process of transforming widescreen films to video formats while preserving the original aspect ratio. For video formats that are almost square this results in masked-off areas above and below the picture area.

LF Line Feed, a control character that commands a printer or other display to move the cursor down one line.

Life Cycle Assessment a process for evaluating the environmental impact of building materials over the entire life of the building including disposal.

Life Cycle Cost total economic cost of a resource, may include cost to create/build, own/operate, and destruction/disposal costs. See also LIFE CYCLE ASSESSMENT.

Life Cycle Cost Calculator financial tool for calculating the economic cost of ownership of a resource or building, see LIFE CYCLE COST.

Life Safety standards and systems involving hazards to human life including fire.

LIFO Last In, First Out, a processing or accounting method where the most recent item is used first, contrast to **FIFO**.

Light-Emitting Diode semiconductor technology used for indicators, lighting, etc.

Light Harvesting see DAYLIGHT HARVESTING.

Light Pipe or **Light Tube,** any cylindrical device used to transport or distribute light; a light pipe may use internal reflection, transmission as in optical fibers, or a combination of both. **DAYLIGHTING** designs use Light Pipes to bring natural light through the roof and into the building interior.

Light Right Consortium an industry and government group promoting a market transformation towards ergonomic lighting through research and tool development; Pacific Northwest National Laboratory is the project manager and Battelle operates the consortium for the US Department of Energy. For more information see www.lightright.org.

Light Shelf as part of a **DAYLIGHTING** schema a white or reflective metal shelf that reflects sunlight upward to illuminate the ceiling may be placed inside or outside of windows on the equator facing side of the structure.

Light-to-Solar Gain (LSG), the ratio between the **SHGC** and **VT** representing the relative efficiency of a glazing material transmitting daylight while blocking heat gains.

Lighting Control System a system that controls electric lights based on a variety of conditions that may include room occupancy, time and external light levels, events, and alarm conditions, etc.

Lights-Out Operation ability to run systems or an entire data center with so few staff that the lights in the machine room can be turned off. Not a power outage.

Limit Switch a protective device used to open or close electrical circuits when limits on temperature, pressure, or mechanical position are reached.

LIMS Laboratory Information Management System.

Link 1) **HTML** shorthand for **HYPERLINK**; 2) a *communications* channel or connection; 3) in *programming*, a technique for combining together separate parts of a larger program or application, also called linkage editing.

Linux a Unix-like computer OPERATING SYSTEM that is OPEN SOURCE and used on a wide variety of devices.

Liquid Crystal Display a display technology that uses transmitted or reflected light, polarizing filters, and a type of semiconductor material that can be electrically switched to allow or block light transmission; also used for electrically controlled privacy glass. For color displays each PIXEL contains three cells (subpixels) with red, green, and blue filters.

Liquid Crystal on Silicon a type of LIQUID CRYSTAL DISPLAY built on a silicon surface with a reflective coating of aluminum. This technology allows smaller pixels and better contrast than transmissive liquid crystal displays.

Listed an attribute of equipment, materials, or services that are included in a list published by an organization concerned with evaluation of products or services whose listing states that either the equipment, material, or service meets appropriate designated standards or has been tested and found suitable for a specified purpose. The listing organization must be acceptable to the AUTHORITY HAVING JURISDICTION and maintain periodic inspection of production of listed equipment or materials or periodic evaluation of services.

LMR or **LMRS,** Land Mobile Radio, short range radios used by companies, local governments, and other organizations to meet a wide range of communication requirements using licensed frequencies below 800 MHz.

Load Balancing 1) *electricity*, distributing load equitably over available circuits and sources; 2) *electrical generation*, the use of various techniques for storing excess power during low demand periods for release during peak demand periods; 3) *computing or networking*, distributing load over multiple computing resources, for example multiple servers supporting the same application.

Load Shedding the ability to turn off or turn down certain equipment in order to reduce the electrical load or demand.

Local Area Network a network within a building or campus. Contrast to **Metropolitan Area Network** or **Wide Area Network**.

Location Based Services wireless information services that are tied to the user's current geographic location. For example, a service that sends you a special offer from a nearby store as you enter a shopping mall.

Locator System or **Location System**, an interactive directory function designed to help people locate other people, meetings, business functions, etc. May be accessible via **Digital Signage**, a **Kiosk**, a **Portal**, or other means.

Lock 1) a *physical* fastening device used to restrict access; 2) *signaling* systems lock onto a clock or other synchronizing signal; 3) in *computer and database* systems a lock or semaphore may be used to limit access to a resource in order to prevent concurrent and inconsistent updates or to prevent changes.

Log 1) a record of events, requests or transactions in time-sequence order; 2) the process of writing such a record; 3) **Log In**, the process of entering identifying and security information required for system access.

Logical Unit Number an identifying number assigned to a device.

LON Local Operating Network, see **LonWorks**.

LonMark LonMark International is a membership organization created to promote open, multi-vendor control systems utilizing ANSI/CEA 709.1, the **LonTalk** standard.

LonTalk a control protocol originally developed by Echelon and formalized as ANSI/CEA 709.1.

LonTalk/IP **LonTalk** routed over IP networks, formalized as CEA-852-A Tunneling Device Area Network Protocols Over Internet Protocol Channels.

LonWorks a control system networking platform created by Echelon and their partners that utilizes ANSI/CEA 709.1 protocols over twisted pair, **POWER LINE CARRIER**, fiber optics or **RF**; there is also an IP tunneling standard, EIA-852.

Look and Feel the general appearance and behavior of a **USER INTERFACE**, including the design, layout, and information content of the display and the elements and actions used to control or manipulate the device.

LOS Line of Sight, electromagnetic radiation (light, microwave, etc.) traveling in a straight line without being blocked by obstructions or the horizon. Contrast to **NLOS**.

Low-e Low Emittance or Low Emissivity, coatings are very thin metal or metallic oxide layers deposited on a window or skylight glazing surfaces to reduce the **U-FACTOR** by suppressing radiative heat flow. Different types of coatings are available for high solar gain, moderate solar gain, or low solar gain.

Low Voltage circuits operating at 50 volts or less (per **NATIONAL ELECTRICAL CODE** article 720); used as shorthand for building related electronics. There are five common low voltage systems (telecommunications or office automations, HVAC, security, fire and life safety, and energy/lighting management) in most buildings, the average commercial office building has 15 low voltage systems and the average hospital has 32 low voltage systems.

lpw (Lumens Per Watt) a measure of the energy efficiency of a light source. Incandescent lamps produce 12-15 lpw; a **COMPACT FLUORESCENT LAMP** produces at least 50 lpw.

LSG (Light-to-Solar Gain), the ratio between the **SHGC** and **VT** representing the relative efficiency of a glazing material transmitting daylight while blocking heat gains.

LSI Large Scale Integration.

LSOH Low-Smoke, zero-Halogen, a plastic used for cable insulation that does not generate the same toxic smoke as **PVC**.

LTE see **3GPP LTE**.

LTL Less Than Truckload.

Lumen the **SI** unit of luminous flux or light intensity. A 100 watt incandescent light bulb emits approximately 1700 lumens.

Lumens Per Watt a measure of the energy efficiency of a light source. Incandescent lamps produce 12-15 lpw; a COMPACT FLUORESCENT LAMP produces at least 50 lpw.

Luminaire any type of lighting fixture or lamp.

LUN (**Logical Unit Number**) an identifying number assigned to a device.

M

M&V Measurement and Verification or Monitoring and Verification, systems and processes for measuring resources usage and verifying any changes in utility charges or savings in energy or water usage.

M13 Mux or **M13 Multiplexer**, a communications device that consolidates multiple **T1** or **E1** links onto one higher speed T3 or DS3 connection (45 Mbit/s) or E3 connection (34 Mbit/s).

M2M (**Machine-To-Machine**) general term for technology that integrates applications using communications interfaces between computers.

MA Mixed Air.

MAC MEDIA ACCESS CONTROL, MESSAGE AUTHENTICATION CODE or MOVES, ADDS AND CHANGES.

MAC Address MEDIA ACCESS CONTROL Address, a (theoretically) unique identifier assigned to each piece of network equipment such as a NETWORK INTERFACE CARD.

Mac OS the operating system (OS) for Apple Macintosh computers. Technically Mac OS is the original version and Mac OS X is the newer UNIX-like operating system.

Machine-To-Machine general term for technology that integrates applications using communications interfaces between computers.

Mail Transfer Agent a computer application or software agent that transfers electronic mail messages from one computer to another.

Main Cross-Connect a STRUCTURED CABLING SYSTEM connection point (see CROSS-CONNECT) between entrance cables, equipment cables, inter-building backbone cables, and intra-building backbone cables of the CORE network.

Main Distribution Area (MDA), an area in a data center that contains the MAIN CROSS-CONNECT and core routers and switches for the LAN, SAN, etc.

Maintenance work performed to ensure proper operation of equipment and facilities including, but not limited to, repair, replacement, and service.

MAN (**Metropolitan Area Network**) a network spanning a campus or city. Contrast to LOCAL AREA NETWORK or WIDE AREA NETWORK.

Management Portal a PORTAL designed to give financial and operational managers access to the facility information they need including energy consumption, utility rates, energy management, occupant information, operational information, and security management information.

MANet Mobile Ad-hoc Network, a WIRELESS AD-HOC NETWORK designed to support mobile nodes and dynamically changing network topology, typically based on certain assumptions about speed and range of movement. See also **VANet**.

Manual Fire Alarm Box a manually operated device used to initiate an alarm signal.

Manufacturing Energy Consumption Survey a **DOE** energy usage survey conducted every four years. For more information see www.eia.doe.gov/emeu/mecs/.

Manufacturing Message Specification ISO 9506 standard for control networks based on a reduced OSI stack with TCP/IP transport/network layer protocol and Ethernet or RS-232C physical media.

MAS Main Automation Supplier.

Mashup a web application that combines content from more than one source. For example, overlaying the locations of jobs from a job board on a map of the area from another site.

Mask temporarily hiding or suppressing an alarm or event.

Master Data Management or **Reference Data Management,** business processes and computer applications designed to manage the consistent use of shared data elements by multiple IT systems and groups.

MasterFormat a CONSTRUCTION SPECIFICATIONS INSTITUTE standard for organizing specifications and other written information for commercial and institutional building projects in the U.S. and Canada. For more information see www.csinet.org.

MAT Mixed Air Temperature.

Material Safety Data Sheet a form describing the properties of a particular substance to provide workers and emergency personnel with safe procedures for handling or working with that substance. Information may include physical data (melting point, boiling point, flash point, etc.), toxicity, health effects, first aid, reactivity, storage, disposal, protective equipment, and spill handling procedures.

MCC or **MC, Main Cross-Connect.**

MCU Micro Control Unit, a control unit containing a microprocessor.

MD5 Message-Digest algorithm 5 or RFC 1321, a cryptographic hash function that creates a 128-bit hash value used to check the integrity of data.

MDA (Main Distribution Area), an area in a data center that contains the **Main Cross-Connect** and core routers and switches for the **LAN, SAN,** etc.

MDF Main Distribution Frame, telephone central office equivalent of **MCC.**

MDI Media or Medium Dependent Interface, an **Ethernet** port used to connect network hubs or switches together without requiring the use of a null-modem or cross-over cable, also called an uplink port.

MDM (Master Data Management) or **Reference Data Management,** business processes and computer applications designed to manage the consistent use of shared data elements by multiple IT systems and groups.

Mechanical Room a space within a building used to house equipment for the mechanical systems such as HVAC, plumbing, and electricity and controllers for the **Building Automation System.**

MECS (Manufacturing Energy Consumption Survey) a **DOE** energy usage survey conducted every four years. For more information see www.eia.doe.gov/emeu/mecs/.

Media Access Control a sub-layer of the data link layer within the **OSI Model** that manages access to a shared physical media.

Media Server a device that stores and shares multimedia content ranging from a media center PC to commercial web servers for large web sites. May include support for STREAMING MEDIA or DIGITAL RIGHTS MANAGEMENT.

Megabit a data transmission speed of one million bits per second. See Fig. 2 on page 279.

Megabyte one million, or ten to the sixth power, bytes (characters) of storage. See Fig. 1 on page 279.

MEMS Micro-Electro-Mechanical Systems.

MEP Mechanical (heating, ventilating, and air conditioning) Electrical and Plumbing.

MERV (Minimum Efficiency Reporting Value), a filter classification based on an **ASHRAE** standard test procedure; the higher the MERV the better the filter. A MERV rating of 15 or higher will trap most biological agents.

MES Manufacturing Execution System.

Mesh Network a type of self-healing network that provides automatic routing of messages between network nodes including the ability to route around blocked paths and the ability to route messages from node to node until a connection can be established. Mesh networks may be self-configuring (so called ad-hoc networks) and may be fixed or mobile (see **MANet**).

Message Authentication Code a short piece of information used to authenticate a message that is calculated using a secret key. A receiver that possesses the secret key can recalculate and check the MAC value to verify the integrity and authenticity of the message.

Metadata data about data, or data that describes a given set of data. For example, metadata may include information on the units used, data format, source, time period, owner, etc.

Metropolitan Area Network a network spanning a campus or city. Contrast to **Local Area Network** or **Wide Area Network**.

Metropolitan Statistical Area (MSA), a geographic areas defined by the US Office of Management and Budget based on Census Bureau data. MSA definitions and associated statistical information are used for a wide variety of purposes.

MF04 shorthand for **MasterFormat** 2004.

MIB Management Information Base, configuration database for **SNMP**.

Microcell a cell in a cellular phone network served by a low power base station covering a limited geographic area such as a shopping mall, hotel, or transportation hub.

Microgrid see **Distributed Energy Resources**.

MID Mobile Internet Device.

Middleware software that enables two or more applications to communicate and exchange data.

MIME (Multipurpose Internet Mail Extensions), an **IETF** standard for extended e-mail message content that supports text and header information in other character sets (not just **ASCII**, non-text attachments, and multi-part message bodies. MIME content type definitions and formats are also used for many other purposes.

Minimum Efficiency Reporting Value (MERV), a filter classification based on an **ASHRAE** standard test procedure; the higher the MERV the better the filter. A MERV rating of 15 or higher will trap most biological agents.

MIPS Millions of Instructions Per Second.

MIS Management Information System.

Mixin an object-oriented programming (see **OO**) technique for implementing INHERITANCE where decisions about how to implement object interface methods are made at the time when the program is run (or runtime) instead of when the program is compiled.

MLC Multi-Loop Controller.

MLS (**Multiple Listing Service**) or **Multiple Listing System**, a database that enables real estate brokers representing sellers to share information about properties with brokers who represent potential buyers or wish to cooperate with a seller's broker in finding a buyer for the property. Listing a property on a service of this type typically implies an offer of compensation to other participants for their services in the sale or lease.

MM Fiber (**Multi-Mode Fiber**), a type of optical fiber that can carry large amounts of power in the 850 nm transmission window over distances of 550 meters or less. Multimode fibers are typically used for premises cabling because they can be powered by LEDs and lower cost VERTICAL CAVITY SURFACE EMITTING LASERs and are easier to splice than the SINGLE MODE FIBER use for long distances.

MMI Man Machine Interface (see USER INTERFACE) or Machine Machine Interface see MACHINE-TO-MACHINE.

MMS MANUFACTURING MESSAGE SPECIFICATION or MULTIMEDIA MESSAGING SERVICE.

Mobile WiMax an enhanced version of **WiMax** that allows users to access the network while moving, standardized as IEEE 802.16e.

MOC Management of Change, see CHANGE MANAGEMENT.

Modbus a family of master-slave control system protocols that can be used on serial links, Ethernet or **TCP/IP** networks. See www.modbus.org for more information.

Monitor a video display device.

MOSFET Metal Oxide Semiconductor Field Effect Transistor.

Motion Pictures Expert Group a group that develops audio and video encoding and compression standards, also used as shorthand for these standards. See www.mpeg.org for more information.

MOU Memorandum of Understanding.

Moves, Adds and Changes the business process for handling the communications network and facilities changes required by user turnover and movement.

MOWS Management Of Web Services, part of the **WSDM** effort that defines the management model for Web Services resources and how to access that manageability using **MUWS**. See www.oasis-open.org for more information.

MP3 a standard for compressed audio recording codified by **Motion Pictures Expert Group** for Layer III of MPEG-1.

MP3 Player any device with the ability to playback **MP3** format audio; may be a special purpose device or a function provided as part of another device such as a telephone or alarm clock.

MPC Multivariable Predictive Control.

MPEG (Motion Pictures Expert Group) a group that develops audio and video encoding and compression standards, also used as shorthand for these standards. See www.mpeg.org for more information.

MPLS (MultiProtocol Label Switching), a networking technology that provides the ability to carry several types of network traffic (multiple protocols) over a shared high speed IP network with enhanced reliability and **availability**; it can also be used to implement **Virtual Private Networks**. MPLS is primarily used by telecommunications carriers and in multi-location enterprise networks.

MPOE Minimum Point Of Entry, termination point for carrier lines, typically 12 inches inside the building foundation.

MPS Master Production Schedule.

MQ 1) IBM WebSphere MQ (formerly MQseries) message queuing MIDDLEWARE products and tools that enable enterprise integration based on reliable asynchronous message passing between applications; 2) similar offerings from other vendors or open source, for example AMQP (www.amqp.org), Apache ActiveMQ (activemq.apache.org), OpenMQ (mq.dev.java.net), JMS, etc.

MR Materials and Resources.

MRL Machine Room Less elevator systems, elevators that do not require a separate space for equipment.

MRO Maintenance, Repair & Operations.

MRP Manufacturing Resource Planning or Materials Resource Planning.

MS/TP Master Slave/Token Passing, a physical networking technology based on **EIA-485** cabling.

MSA (Metropolitan Statistical Area), a geographic areas defined by the US Office of Management and Budget based on Census Bureau data. MSA definitions and associated statistical information are used for a wide variety of purposes.

MSDS (Material Safety Data Sheet) a form describing the properties of a particular substance to provide workers and emergency personnel with safe procedures for handling or working with that substance. Information may include physical data (melting point, boiling point, flash point, etc.), toxicity, health effects, first aid, reactivity, storage, disposal, protective equipment, and spill handling procedures.

MSG Model Support Group of the INTERNATIONAL ALLIANCE FOR INTEROPERABILITY.

MTA (Mail Transfer Agent) a computer application or software agent that transfers electronic mail messages from one computer to another.

MTBF Mean Time Between Failures, the average time between device or system failures.

MTO Make to Order.

MTTR Mean Time To Restore (or repair), average time required to return a failed device or system to service.

Multi-Mode Fiber (MM Fiber), a type of optical fiber that can carry large amounts of power in the 850 nm transmission window over distances of 550 meters or less. Multimode fibers are typically used for premises cabling because they can be powered by LEDs and lower cost VERTICAL CAVITY SURFACE EMITTING LASERs and are easier to splice than the SINGLE MODE FIBER use for long distances.

Multicast transmitting information to a group of destinations simultaneously. Efficient multicast within a data network requires special routing software designed to deliver the messages over each link of the network only once and create copies only at the points where the paths to the destinations separate.

Multicasting a technique for selective many-to-many network communication. IP Multicast uses specially designated group addresses in the range 224.0.0.0 to 239.255.255.255; the sender sends a single datagram to the multicast address and the routers take care of making copies and sending them to all receivers that have registered their interest in data from that sender.

Multimedia combining more than one media such as audio with still or moving pictures.

Multimedia Messaging Service (MMS), an extended form of SHORT MESSAGE SERVICE that supports multimedia objects (images, audio, video, rich text) and not just text.

Multiple Listing Service or **Multiple Listing System,** a database that enables real estate brokers representing sellers to share information about properties with brokers who represent potential buyers or wish to cooperate with a seller's broker in finding a buyer for the property. Listing a property on a service of this type typically implies an offer of compensation to other participants for their services in the sale or lease.

Multiplexing techniques for simultaneous or sequential transmission of multiple signals on one circuit or communications channel, including means for positively identifying each signal.

MultiProtocol Label Switching (MPLS), a networking technology that provides the ability to carry several types of network traffic (multiple protocols) over a shared high speed IP network with enhanced reliability and AVAILABILITY; it can also be used to implement VIRTUAL PRIVATE NETWORKS. MPLS is primarily used by telecommunications carriers and in multi-location enterprise networks.

Multipurpose Internet Mail Extensions (MIME), an **IETF** standard for extended e-mail message content that supports text and header information in other character sets (not just **ASCII,** non-text attachments, and multi-part message bodies. MIME content type definitions and formats are also used for many other purposes.

MUTOA Multi-User Telecommunications Outlet Assembly, an end user connecting point in a STRUCTURED CABLING SYSTEM; technically a form of work area outlet.

MUWS Management Using Web Services, part of the **WSDM** effort that defines Web Services interfaces for management providers. See www.oasis-open.org for more information.

MVC Multivariable Control.

MVS Machine Vision System

MX Record Mail eXchanger record, a specialized type of **DOMAIN NAME SERVER** record used to control the routing of e-mail messages.

N

N+1 a design for increased reliability that includes one additional unit and enables the additional unit to take over for any other unit in the event of an outage.

NAFEM (North American Association of Food Equipment Manufacturers) a trade organization of foodservice equipment and supplies manufacturers. See www.nafem.org for more information.

NAFEM Data Protocol standard rules and format for communications between commercial kitchen equipment and foodservice management systems covering these critical areas: product/inventory management, asset management, labor management, food safety, and energy management. See www.nafem.org for more information.

NAICS North American Industry Classification System, coding scheme used to classify economic activity, designed to replace **SIC** and be consistent in the US, Mexico, and Canada. The first two digits designate the business sector, the third digit designates the subsector, the fourth digit designates the industry group, and the fifth digit designates particular industries.

NAK Negative Acknowledgement, a character or protocol message indicating an error condition. Contrast to **ACKNOWLEDGE**.

NAS NETWORK ACCESS SERVER or **NETWORK ATTACHED STORAGE**.

NAT (Network Address Translation) a function provided by a **ROUTER** where device addresses are mapped or translated between network segments.

National Building Information Model Standard a common integrated life-cycle information model for the **A/E/C** and **FM** industries being developed by a committee of the **NATIONAL INSTITUTE OF BUILDING SCIENCES FACILITY INFORMATION COUNCIL**. For more information see www.facilityinformationcouncil.org/bim/.

National Electrical Code the US code for electrical fire safety published by the **NFPA**; see www.nfpa.org for more information.

National Fenestration Rating Council (NFRC), a nonprofit organization that rates the energy performance of windows, doors, skylights, and attachment products. For more information see www.nfrc.org.

National Fire Protection Association codes and standards organization for fire, electrical and building safety, see www.nfpa.org for more information.

National Institute of Building Sciences a non-profit, non-governmental organization established by the U.S. Congress to serve as an interface between government and the private sector in order to improve the building regulatory environment; facilitate the introduction of new technology into the building process; and to disseminate technical and regulatory information. For more information see www.nibs.org.

National Pollutant Discharge Elimination System the US **EPA** system for granting and regulating point and non-point sources that discharge pollutants into waters of the United States. For more information see cfpub.epa.gov/npdes/.

NBIMS (National Building Information Model Standard) a common integrated life-cycle information model for the A/E/C and **FM** industries being developed by a committee of the NATIONAL INSTITUTE OF BUILDING SCIENCES FACILITY INFORMATION COUNCIL. For more information see www.facilityinformationcouncil.org/bim/.

NC NETWORK COMPUTER or Numerical Control see COMPUTER NUMERICAL CONTROL.

NDS see VIDEOGUARD.

Near Field Communications (NFC), a short-range wireless technology standardized as ECMA-340 and ISO/IEC 18092 that uses the unlicensed RF band at 13.56 MHz to deliver 106 kbit/s, 212 kbit/s or 424 kbit/s at distances of up to 0.2m. This technology is currently used with mobile phones for mobile commerce, keys, identity, tickets, etc. For more information see www.nfc-forum.org.

NEC US NATIONAL ELECTRICAL CODE or NEC Corporation.

NEMA National Electrical Manufacturers' Association, see www.nema.org for more information.

Network any system of interconnected devices or processes including telecommunications networks, business networks (also known as supply chains or value networks), social networks, etc.

Network Access Server a network security device that controls access to a network and resources on the network.

Network Address Translation a function provided by a ROUTER where device addresses are mapped or translated between network segments.

Network Administrator a person or group that maintains network infrastructure such as switches, routers, cabling, and telecommunications carrier connections; diagnoses network and communications problems; administers network addresses and configuration information; and with these or with the behavior of network-attached computers.

Network Attached Storage storage systems that can be accessed over a network using file level protocols. Similar to **SAN**.

Network Computer a limited function computer system that operates via a network connection and does not have a local hard disk drive.

Network Interface Card the physical interface between a device and the network. The use of the term card is a historical artifact; this function may be a single chip or embedded in a larger device.

Network Operations Center a location or function for monitoring and managing the operation of one or more communications networks and associated equipment.

Networked Virtual Ecosystem (NVE), the Value Network of organizations that cooperate to deliver products and services to the customer of a NETWORKED VIRTUAL ORGANIZATION.

Networked Virtual Organization (NVO), a strategy for organizational transformation based on improving the end-customer experience, focusing internal operations on key value creating functions and outsourcing tasks that do not differentiate the organization, and standardizing business operations, data, and information technology in order to increase efficiency.

Neuron chip an interface chip for LonWorks networks that provides LonTalk protocol support, a unique node identifier, and other functionality.

NFC (Near Field Communications), a short-range wireless technology standardized as ECMA-340 and ISO/IEC 18092 that uses the unlicensed RF band at 13.56 MHz to deliver 106 kbit/s, 212 kbit/s or 424 kbit/s at distances of up to 0.2m. This technology is currently used with mobile phones for mobile commerce, keys, identity, tickets, etc. For more information see www.nfc-forum.org.

NFPA (National Fire Protection Association) codes and standards organization for fire, electrical and building safety, see www.nfpa.org for more information.

NFRC (National Fenestration Rating Council), a nonprofit organization that rates the energy performance of windows, doors, skylights, and attachment products. For more information see www.nfrc.org.

NIBS (National Institute of Building Sciences) a non-profit, non-governmental organization established by the U.S. Congress to serve as an interface between government and the private sector in order to improve the building regulatory environment; facilitate the introduction of new technology into the building process; and to disseminate technical and regulatory information. For more information see www.nibs.org.

NIC (Network Interface Card) the physical interface between a device and the network. The use of the term card is a historical artifact; this function may be a single chip or embedded in a larger device.

NIMBY Not In My Back Yard, a shorthand term describing political resistance to building or construction projects.

NIST US National Institute of Standards and Technology, see www.nist.gov.

NLOS (Non Line Of Sight) some technologies like microwave are limited to line of sight transmission; this term refers to technologies that are not subject to this limitation.

NML Network Management Layer.

NNTP Network News Transfer Protocol, an Internet application protocol for reading and posting USENET articles, as well as transferring news among news servers.

NOC (Network Operations Center) a location or function for monitoring and managing the operation of one or more communications networks and associated equipment.

Node any device that is connected as part of a communications network.

NOI Net Operating Income.

Non Line Of Sight some technologies like microwave are limited to line of sight transmission; this term refers to technologies that are not subject to this limitation.

Normal Power Factor Ballast a BALLAST with a POWER FACTOR between 0.4 and 0.6.

North American Association of Food Equipment Manufacturers a trade organization of foodservice equipment and supplies manufacturers. See www.nafem.org for more information.

Nosocomial Infections hospital acquired infections secondary to the patient's original condition.

Notification Appliance a fire alarm system component that provides audible, tactile, or visible outputs. For example, a bell, horn, loud speaker, strobe light, or text display.

NPDES (National Pollutant Discharge Elimination System) the US EPA system for granting and regulating point and non-point sources that discharge pollutants into waters of the United States. For more information see cfpub.epa.gov/npdes/.

NPF Ballast (Normal Power Factor Ballast) a BALLAST with a POWER FACTOR between 0.4 and 0.6.

NPV Net Present Value, the present financial value of a series of future cash flows.

NTE National Transportation Exchange.

NTP Network Time Protocol, see also **SNTP**.

NTSC the analog television standard used in North America, Japan, and certain other areas.

Nuisance Alarm an alarm that is caused by a malfunction or spurious trigger.

Null 1) *general* usage is nothing or zero; 2) *null character* is all zero bits and may be used as a string terminator or filler; 3) *null modem* is a crossover cable used to connect two devices without actual modems; 4) *null value* for a field in a RELATIONAL DATABASE MANAGEMENT SYSTEM indicates a missing or unknown value and is different from an empty value.

NURBS Non-Uniform Rational B Splines, mathematical formulas used to create computer models of curves and surfaces.

NVE (Networked Virtual Ecosystem), the Value Network of organizations that cooperate to deliver products and services to the customer of a NETWORKED VIRTUAL ORGANIZATION.

NVO (Networked Virtual Organization), a strategy for organizational transformation based on improving the end-customer experience, focusing internal operations on key value creating functions and outsourcing tasks that do not differentiate the organization, and standardizing business operations, data, and information technology in order to increase efficiency.

NVR Network Video Recorder, a video storage system designed to attach to network and support IP cameras.

O

O&M Operations and Maintenance.

OA (Outside Air) fresh air from outside the building.

OA&M Operation, Administration and Maintenance.

OAM Operation, Administration, and Maintenance.

OAS Outside Air Supply.

OASIS (Organization for the Advancement of Structured Information Standards), is a global consortium that facilitates the development, convergence and adoption of e-business and web service standards. For more information see www.oasis-open.org.

OAT Outdoor Air Temperature.

OBEX OBject EXchange, a communications protocol that enables the exchange of binary objects such as business cards, data, and applications between devices such as **PDA**s and mobile phones via **INFRARED** or **BLUETOOTH**. See www.irda.org for more information.

oBix open Building Information eXchange, a standard for **WEB SERVICES** based protocols that enable communications between building mechanical and electrical systems and enterprise applications. For more information see www.obix.org.

Object in object-oriented programming (see **OBJECT-ORIENTED**) an object is an instance or a particular example of a class that encapsulates the program code and data used to represent some kind of thing in the real world.

Object Database Management System software for storing and retrieving information that is specifically designed for high performance processing of complex data using **OBJECT-ORIENTED** programming techniques and languages.

Object-Oriented an approach to computer programming where programs are composed of objects (rather than functions or procedures) and each object encapsulates logic and data that can be manipulated by a defined set of interface methods or messages. The modifier object-oriented can be applied to analysis, design, programming, programming languages, databases, etc.

OC-n Optical Carrier-n, data rates for **SONET** networks. The most common rates are OC-1 is 51.84 Mbit/s, OC-3 is 155.52 Mbit/s, and OC-12 is 622.08 Mbit/s.

OCC Old Corrugated Cardboard for recycling or **OPERATIONS CONTROL CENTER**.

OCCS **OMNICLASS** Construction Classification System.

Occupancy Management may be either a system for managing an organization's facilities and obligations or a system for tracking the presence of people in various areas of a facility using OCCUPANCY SENSORS and other techniques.

Occupancy Pushbutton a manual occupancy indicator.

Occupancy Sensor automatic technology to sense movement or other parameters in order to determine if a space is occupied for lighting control, HVAC, security and other applications.

Occupant Portal a PORTAL designed to give building occupants access to the information they need about their space, to control their comfort parameters, and to enter facilities requests. May include other functions like monitoring energy usage, guest management, parking, etc.

OCR (Optical Character Recognition) the ability to process image data and recognize letters and numbers contained in the image. For example, the ability to capture text by scanning and processing a paper document or the ability to read license plate numbers from a photo.

OCS (Open Control System) control systems that use OPEN PROTOCOLS and OPEN STANDARDS. Not to be confused with open loop control, without feedback.

OCX OLE Custom Control, a programmatic extension to Microsoft OLE.

ODBC (Open DataBase Connectivity), a standard APPLICATION PROGRAMMING INTERFACE for accessing data from a RELATIONAL DATABASE MANAGEMENT SYSTEM using the STRUCTURED QUERY LANGUAGE that can also be used to access spreadsheets, text files containing tabular data, and XML files. ODBC can be used within a single system or between systems.

ODBMS (Object Database Management System) software for storing and retrieving information that is specifically designed for high performance processing of complex data using OBJECT-ORIENTED programming techniques and languages.

ODP (Ozone-Depletion Potential), a measure of the potential impact on the ozone layer of a chemical; ODP is expressed as a ratio relative to the impact of a similar mass of R11 or CFC-11. See Fig. 3 on page 280 for example values.

ODS Ozone-Depleting Substance.

OEM Original Equipment Manufacturer.

OFAC or **Office of Foreign Assets Control,** a US Treasury function that administers economic and trade sanctions based on foreign policy and national security goals. US businesses may be required to check their overseas trading partners against OFAC Specially Designated National lists and report suspicious transactions.

OFCI Owner Furnished and Contractor Installed.

OFE Owner Furnished Equipment.

OFOI Owner Furnished and Owner Installed.

OGC (Open Geospatial Consortium) an international organization that is leading the development of standards for geospatial and location based services. For more information see www.opengeospatial.org.

OH (Overhead) 1) *business* costs that are not directly related to revenue generation; 2) *systems*, non-productive resource requirements or usage; 3) *communications* additional data, processing, or time required to support a protocol or transmission media; 4) *HVAC* systems that deliver and remove air from above, contrast to **UFAD**.

OI Operator Interface.

OLAP Online Analytical Processing, systems and specialized databases designed to provide fast answers to analytical queries using aggregated information.

OLE Object Linking and Embedding, a distributed object system and protocol developed by Microsoft primarily for creating compound documents. Replaced by **COM** and **DCOM**.

OLED (Organic Light-Emitting Diode) a type of LIGHT-EMITTING DIODE where the emissive layer is a thin-film of certain organic compounds.

OLTP (Online Transaction Processing), a type of system designed to reliably process business transactions, frequently in high volumes and with high reliability requirements. Contrast to **OLAP**, see also TRANSACTION.

OMA (Open Mobile Alliance), an organization of mobile operators, device and network suppliers, information technology companies and content and service providers working to develop standards for mobile services. See www.openmobilealliance.org for more information.

OMAC Open Modular Architecture Control, manufacturing users group, see www.omac.org for more information.

OMG Object Management Group, a computer industry specifications consortium. See www.omg.org for more information.

OmniClass or **OmniClass Construction Classification System,** a new construction industry classification system that incorporates existing systems including **MASTERFORMAT** for work results, **UNIFORMAT** for elements, and **EPIC** for structuring products. For more information see www.omniclass.org.

OMS On-line Management System.

On-Demand access to information or resources based on user request as contrasted to pre-scheduled access or broadcast.

Online Transaction Processing (OLTP), a type of system designed to reliably process business transactions, frequently in high volumes and with high reliability requirements. Contrast to **OLAP**, see also **Transaction**.

OO Object-Oriented or Owners and Operators of buildings.

OOH (Out-Of-Home), advertising media that reaches the consumer while outside the home including **digital signage**, billboards, street furniture, transit advertising, etc. Contrast to broadcast, print, or Internet media that are usually for home or office viewing.

OPC (Openness, Productivity, Connectivity Foundation) (formerly OLE for Process Control), an automation industry group that develops protocol standards; their original standards were Microsoft oriented and their current standards are based on **Web Services**. See www.opcfoundation.org for more information.

OPC Foundation Unified Architecture a robust and scalable platform for all OPC applications that is based on TCP/IP, HTTP, SOAP, and XML instead of Microsoft **COM/DCOM**. See www.opcfoundation.org for more information.

OPC UA (OPC Foundation Unified Architecture) a robust and scalable platform for all OPC applications that is based on TCP/IP, HTTP, SOAP, and XML instead of Microsoft **COM/DCOM**. See www.opcfoundation.org for more information.

OPC XMLDA or **OPC XML-DA**, an industrial automation data access protocol based on **Web Services** standards; standardizing the messages exchanged instead of the **API** allows implementation on different operating systems. See www.opcfoundation.org for more information.

Open Control System control systems that use **open protocol**s and **open standard**s. Not to be confused with open loop control, without feedback.

Open Database Connectivity (ODBC), a standard APPLICATION PROGRAMMING INTERFACE for accessing data from a RELATIONAL DATABASE MANAGEMENT SYSTEM using the STRUCTURED QUERY LANGUAGE that can also be used to access spreadsheets, text files containing tabular data, and **XML** files. ODBC can be used within a single system or between systems.

Open Design Alliance (formerly OpenDWG Alliance), an association of software developers and users promoting open, industry-standard formats for the exchange of CAD data such as OPENDWG and OPENDGN. For more information see www.opendesign.com.

Open Geospatial Consortium an international organization that is leading the development of standards for geospatial and location based services. For more information see www.opengeospatial.org.

Open Mobile Alliance (OMA), an organization of mobile operators, device and network suppliers, information technology companies and content and service providers working to develop standards for mobile services. See www.openmobilealliance.org for more information.

Open Protocol a protocol that is documented in publicly available standards that may be implemented by anyone.

Open Source software provided under terms that include access to the source code and specific license terms that allow improvement and distribution.

Open Standard publicly available standards that may be implemented by anyone.

Open Standards Consortium for Real Estate a group developing standards and XML schemas for transferring real estate information between systems. For more information see www.oscre.org.

Open Systems computer systems that are interoperable, portable, and free from proprietary standards.

Open Systems Integration and Performance Standards a series of data models being defined by the **SIA** to improve interoperability between components in access control systems. For more information see www.siaonline.org.

OpenDGN an open **CAD** file specification based on information provided by Bentley Systems; the **OPEN DESIGN ALLIANCE** offers software library support.

OpenDWG a **CAD** file format developed by the **OPEN DESIGN ALLIANCE** as a fully interchangeable version of the Autodesk **DWG** format.

Openness, Productivity, Connectivity Foundation (formerly OLE for Process Control), an automation industry group that develops protocol standards; their original standards were Microsoft oriented and their current standards are based on **WEB SERVICES**. See www.opcfoundation.org for more information.

Operating System a program that manages hardware and software resources within a computer or a microprocessor based device. For example Linux, Mac OS, Microsoft Windows, Palm OS, etc.

Operations Control Center a location or function for monitoring and managing the operation of some group of systems for facilities; for example building operations, information systems operations, etc. See also **NETWORK OPERATIONS CENTER**.

Operations Interface the **USER INTERFACE** provided for operations personnel.

Operator Workstation the user interface to control systems.

Opex or **Operating Expense**, expenses incurred as part of operations and having a useful life of less than one accounting period. Operating expenses are not subject to depreciation or amortization; contrast to **CAPEX**.

Optical Character Recognition the ability to process image data and recognize letters and numbers contained in the image. For example, the ability to capture text by scanning and processing a paper document or the ability to read license plate numbers from a photo.

Organic Light-Emitting Diode a type of LIGHT-EMITTING DIODE where the emissive layer is a thin-film of certain organic compounds.

Organization for the Advancement of Structured Information Standards (OASIS), is a global consortium that facilitates the development, convergence and adoption of e-business and web service standards. For more information see www.oasis-open.org.

ORP Oxygen-Reduction Potential, a measurement reflecting the concentration of oxidizing agents in water or other fluids.

OS (Operating System) a program that manages hardware and software resources within a computer or a microprocessor based device. For example Linux, Mac OS, Microsoft Windows, Palm OS, etc.

OSCRE (Open Standards Consortium for Real Estate) a group developing standards and XML schemas for transferring real estate information between systems. For more information see www.oscre.org.

OSF Open Software Foundation, an organization that develops standards for UNIX and related software, now part of THE OPEN GROUP.

OSHA Occupational Safety & Health Administration, a US government agency.

OSI Open Systems Interconnection, an early ISO and ITU effort to standardize data networking protocols.

OSI Model Open Systems Interconnection Model, OSI Reference Model, or ISO 7498, a common way of describing communications and computer network protocols that divides protocols into seven layers, sometimes called a protocol stack. As shown in Fig. 6 on page 282, the layers are (starting from the bottom) Physical, Data Link, Network Layer, Transport, Session, Presentation, and Application.

OSIPS (Open Systems Integration and Performance Standards) a series of data models being defined by the **SIA** to improve interoperability between components in access control systems. For more information see www.siaonline.org.

OSS Operations Support Systems.

OSSTMM Open Source Security Testing Methodology Manual, a peer-reviewed methodology for performing security tests. See www.isecom.org/osstmm/ for more information.

Out-Of-Home (OOH), advertising media that reaches the consumer while outside the home including DIGITAL SIGNAGE, billboards, street furniture, transit advertising, etc. Contrast to broadcast, print, or Internet media that are usually for home or office viewing.

Outlook a Microsoft e-mail client product that supports calendar management and other collaboration features as part of the Office Suite; Outlook Express is a more limited e-mail client provided as part of Microsoft Windows.

Outside Air fresh air from outside the building.

OVAB Out-of-Home Video Advertising Bureau, a membership organization of OUT-OF-HOME video networks and advertising companies that supports planning, buying and evaluating the effectiveness of the medium. For more information see www.ovab.org.

Overhead 1) *business* costs that are not directly related to revenue generation; 2) *systems*, non-productive resource requirements or usage; 3) *communications* additional data, processing, or time required to support a protocol or transmission media; 4) *HVAC* systems that deliver and remove air from above, contrast to **UFAD**.

OWS (Operator Workstation) the user interface to control systems.

Ozone-Depletion Potential (ODP), a measure of the potential impact on the ozone layer of a chemical; ODP is expressed as a ratio relative to the impact of a similar mass of R11 or CFC-11. See Fig. 3 on page 280 for example values.

P&ID Process and Instrumentation Diagram.

P&L Profit and Loss.

P2P (Peer To Peer) a networking technique for communication between users (peers) without a central system or server.

Packet Switching a type of computer and telecommunications networking technology based on routing packets (units of information) over shared data links between network nodes. Contrast to CIRCUIT SWITCHING where links are not shared.

PACS PHYSICAL ACCESS CONTROL SYSTEM or PICTURE ARCHIVING AND COMMUNICATION SYSTEMS.

PAH Polycyclic Aromatic Hydrocarbons, a class of chemical compounds found in oil and tar and also formed during the incomplete burning of coal, oil, gas, wood, garbage, tobacco, fat, or other organic substances. These compounds are considered to be organic pollutants; some are known or suspected carcinogens and linked to other health problems.

PAL or **Phase Alternating Line**, a 625-line 50 Hz broadcast television standard used primarily in Europe. Also known as 576i.

Pan to move a camera or other device from side to side.

PAN PERSONAL AREA NETWORK see also **WPAN** or Primary Account Number.

Parity or **Parity Bit,** redundant information that is added and used for error detection.

Parking Ratio number of parking spaces compared to the size of the building, may be expressed in spaces per 1,000 SF of GBA (Gross Building Area) or GLA (Gross Leasable Area), etc.

PAS Process Automation System.

Passive IR or **Passive Infrared,** technology used for motion detectors and other applications that work based on changes in reflected infrared light.

Passive RFID a **RFID** device that does not have an internal power source and only transmits signals in the presence of an incoming radio signal that provides both power and information. For example, some types of **Smart Card** or **Fob System** devices, tags that can be attached to packages or equipment, or chips that can be implanted in animals.

Passive Solar a general class of technologies that convert sunlight into usable heat, cause air-movement for ventilation or cooling, or store heat for future use, without using other energy sources to power pumps or fans. Contrast to **Active Solar.**

Patch Cable or **Patch Cord,** any electrical cable for signal routing used to connect ("patch-in") one electronic device to another. Patch cables are typically short and used with a **Patch Panel.**

Patch Panel a panel with a number of jacks for labeling, monitoring, interconnecting, and testing circuits in a convenient, flexible manner. Typically short **patch cables** will plug into the front side while the back is connected to the longer and more permanent cables.

Payment Card Industry (PCI), 1) *general* term for the payment industry including debit, credit, pre-paid, e-purse, automated teller machines, POS, etc.; 2) *PCI Security Standards Council,* an industry standards development organization, see www.pcisecuritystandards.org and **PCI DSS.**

PB (**Petabyte**) one quadrillion, or ten to the fifteenth power, bytes (characters) of storage. See Fig. 1 on page 279.

PBO (**Plan Build and Operate**) a business model where one organization has responsibility for all phases of a project including planning, construction, and operation.

PBX (**Private Branch eXchange**) a telephone switch (or exchange) owned by a private business.

PC Personal Computer.

PCAOB Public Company Accounting Oversight Board, a nonprofit corporation created by SARBANES-OXLEY to oversee public company auditors and establish standards. For more information see www.pcaobus.org.

PCB (**Polychlorinated Biphenyls**) a class of organic compounds used for a variety of industrial applications including transformers and capacitors. These compounds are toxic and persistent organic pollutants; for more information see www.epa.gov/pcb/.

PCI PAYMENT CARD INDUSTRY or PERIPHERAL COMPONENT INTERCONNECT.

PCI DSS Payment Card Industry Data Security Standards, security standards for account data protection developed by the PCI Security Standards Council. For more information see www.pcisecuritystandards.org.

PCM (**Pulse-Code Modulation**) a digital representation of an analog signal where the magnitude of the signal is sampled regularly at uniform intervals.

PCO Proposed Change Order, see CHANGE ORDER.

PCS PERSONAL COMFORT SYSTEM, PERSONAL COMMUNICATIONS SERVICE, or Process Control System.

PD Positive Displacement, see **Positive Displacement Flowmeter** or **Positive Displacement Pump** depending on the context; **Powered Device**.

PDA (Personal Digital Assistant) a handheld device that can be used as personal organizers and may also incorporate other functions including games, music, cameras, wireless voice and data communications, etc.

PDF (Portable Document Format) an open standard file format, proprietary to Adobe Systems, for representing documents in a way that is consistent on different hardware and software.

PDM Project Data Management or Product Data Management.

PDP Plasma Display Panel or Power Distribution Panel.

PDP TV Plasma Display Television.

PDS Plant Design System.

PDU Power Distribution Unit.

PE Premises Equipment or Professional Engineer.

Peak Shaving a **Demand Management** term for reducing loads during periods of high demand.

Peer To Peer a networking technique for communication between users (peers) without a central system or server.

PEG (Personal Entertainment Guide) an online interactive guide to available media geared to an individual viewer, not a household, and potentially covering multiple viewing media and sources.

PEL Permissible Exposure Limit, regulatory limits on the amount or concentration of a substance in the air.

Pen Testing a shorthand term for **Penetration Testing**.

Penetration Testing a method of evaluating the security of a system or network by simulating a hacker attack.

Perimeter Intrusion Detection System an automated system for monitoring the exterior of a building or area.

Peripheral Component Interconnect (PCI), a series of standards for a computer BUS used to attach peripheral devices to a computer motherboard and processor.

Permanent Virtual Circuit (PVC), a fixed connection between two end points in a **VLAN, VPN** or other network.

Personal Area Network a network connecting devices close to one person such as telephones and **PDAs**. Wired personal area network technologies include **USB** and **FIREWIRE**; wireless technologies include **BLUETOOTH** and infrared (**IrDA**).

Personal Comfort System the ability for users to control the temperature and other environmental parameters in their space.

Personal Communications Service cellular telephone service using a variety of signaling standards and the 1900-MHz radio band in North America or 1800-MHz band in Europe and Asia.

Personal Digital Assistant a handheld device that can be used as personal organizers and may also incorporate other functions including games, music, cameras, wireless voice and data communications, etc.

Personal Entertainment Guide an online interactive guide to available media geared to an individual viewer, not a household, and potentially covering multiple viewing media and sources.

Personal Identity Verification (PIV), 1) *general*, a security application for identity and CREDENTIAL management; 2) *specific* cards and identity management services that conform to the requirements of **FIPS 201**.

Personal Lighting Control the ability for an individual to change lighting levels in their space.

Personal Video Recorder (PVR), a consumer appliance or other device such as PC hardware and software, used to record and playback video.

Personalization tailoring something based on the user's personal details, characteristics, or preferences. For example personalizing a SMART CARD, a PORTAL start page, etc.

PES Process Electrochemical Systems.

Petabyte one quadrillion, or ten to the fifteenth power, bytes (characters) of storage. See Fig. 1 on page 279.

PFC Power Factor Correction, systems designed to keep the POWER FACTOR of a load or building near one by switching in or out capacitors or inductors which act to cancel the inductive or capacitive effects of the load.

Phantom Load electrical power used by a device that is switched off.

Phase-Locked Loop a closed-loop feedback control system that generates an output based on the frequency and phase of an input signal; used to stabilize a generated signal or to detect signals in the presence of noise.

Phishing efforts to fraudulently acquire valuable information, such as passwords and credit card numbers, by masquerading as a business or a trustworthy person in electronic communication.

phoneME an OPEN SOURCE project addressing the market and technical requirements of "feature phone" devices using J2ME software. For more information see phoneme.dev.java.net.

Photo Call-Up a security system feature where a stored image is automatically displayed in order to validate the person presenting the credential.

Photovoltaic or **Solar Cells,** materials that convert light into electrical power. See **Solar Panel.**

PHY 1) the physical layer of **OSI model;** 2) the name used for the integrated circuit or functional component that provides physical interface functionality for implementing **Ethernet, USB,** etc.

Physical Access Control System a system for monitoring, controlling, and recording entry and exit to secured areas.

PICS Platform for Internet Content Selection or **Protocol Implementation Conformance Statement.**

Picture Archiving and Communication Systems technology for storing medical images and making them available for distribution and presentation over a network.

PID (Proportional-Integral-Derivative) a type of digital or analog controller or control algorithm that uses a signal, its cumulative value, and rate of change in feedback calculations to adjust the controlled quantity to a given **setpoint** value.

PIDS (Perimeter Intrusion Detection System) an automated system for monitoring the exterior of a building or area.

PIM Plant Information Management.

PIMS Process Information Management System.

PIN Personal Identification Number.

PIP Partner Interface Process.

PIR (Passive IR) or **Passive Infrared,** technology used for motion detectors and other applications that work based on changes in reflected infrared light.

PIV (Personal Identity Verification), 1) *general*, a security application for identity and CREDENTIAL management; 2) *specific* cards and identity management services that conform to the requirements of **FIPS 201**.

Pixel a picture element, the smallest element of data in an image.

PKI (Public Key Infrastructure) a system for trusted third party vetting of, and vouching for, user identities and associating public encryption keys to users based on **X.509** or other standards. Public keys are typically in a DIGITAL CERTIFICATE that contains a DIGITAL SIGNATURE from a CERTIFICATE AUTHORITY.

Placemark in **KML** or Google terminology a placemark can include a location (latitude and longitude), viewpoint information (altitude, direction, angle), and an icon or marker.

Plan Build and Operate a business model where one organization has responsibility for all phases of a project including planning, construction, and operation.

Plasma Display a flat screen display technology that uses tiny cells containing an mixture of noble gases (neon and xenon) that can be electrically excited into a plasma which then excites phosphors to emit light.

Plasma Display Panel another term for a PLASMA DISPLAY.

Platform the supporting systems and facilities for an application or service typically including the hardware, OPERATING SYSTEM, and other resources.

Platform for Internet Content Selection content labeling standard originally designed to help control what children access on the Internet, also used for code signing and privacy. See www.w3.org/PICS/ for more information.

Platinum Resistance Thermometer same as RESISTANCE TEMPERATURE DETECTOR.

PLC Power Line Carrier or Programmable Logic Controller.

Plenum 1) a sealed chamber at the inlet or outlet of an air handler, ducts attach to the plenum; 2) space above a dropped ceiling or below a raised floor sometimes used as a return or supply air path for HVAC and for cabling.

PLL (Phase-Locked Loop) a closed-loop feedback control system that generates an output based on the frequency and phase of an input signal; used to stabilize a generated signal or to detect signals in the presence of noise.

PLMRS or **Private Land Mobile Radio Service**, see **LMR**.

Plug-and-Play technology that enables a system to automatically detect and configure a new device when the device is added. Originally developed for PC devices, the concept has been extended to networks (see Universal Plug and Play) and other systems.

Plug-In or **Plugin**, add-on software that can easily be installed and accessed. For example, most web browsers allow the use of plug-in modules to display animated or video content.

PM Particulate Matter or Project Manager.

PM&C Process Monitoring & Control.

PM10 PM_{10} or **Large Particulates**, airborne particulate matter or aerosols with particle sizes of 10 microns or micrometers or less. Larger particles are filtered in the nose and throat but particulate matter smaller than about 10 micrometers can settle in the bronchi and lungs and cause health problems.

PM2.5 or **$PM_{2.5}$** or **Small Particulates**, airborne particulate matter or aerosols with particle sizes of 2.5 microns or micrometers or less. Larger particles are filtered in the nose and throat but particles this small can penetrate into the gas-exchange regions of the lung.

PMD or **Programmable Message Display**, a type of **Digital Signage**.

Pneumatic Control control systems that use compressed gas, typically pressurized air, for signaling.

PNL Panel.

POE Post-Occupancy Evaluation or **Power over Ethernet**.

Point of Presence a **demarcation point** between communications entities or an access point for a network. Long distance carriers and Internet service providers typically have multiple POPs where customers connect to their network.

Poly Vinyl Chloride (PVC), a thermoplastic polymer used for many purposes including cable insulation and building materials. The plasticized form of PVC used for cable insulation forms toxic smoke when burned, leaded PCV cable jackets breakdown and produce lead dust, and there are environmental concerns about other forms of PVC as well.

Polychlorinated Biphenyls a class of organic compounds used for a variety of industrial applications including transformers and capacitors. These compounds are toxic and persistent organic pollutants; for more information see www.epa.gov/pcb/.

POP Point of Presence or **Post Office Protocol**, depending on the context.

POP3 Post Office Protocol3, an application level protocol for e-mail access.

Port 1) a *hardware* port is a connecting point or jack; 2) a *networking* port is a sub-address that may be associated with a specific application or usage, for example **TCP** protocol associates port 80 with **HTTP** Web access; 3) to *port software* is the process of converting and testing software in a different environment (or a different version of the same platform), for example porting a Linux program to Microsoft Windows.

Portable Document Format an open standard file format, proprietary to Adobe Systems, for representing documents in a way that is consistent on different hardware and software.

Portal 1) a framework or software application that brings together information from multiple systems in a single user interface; 2) a website designed to serve as an entry-point to many other sites.

POS Point of Sale.

POSIT or **Profiles for Open Systems Internetworking Technologies**, US standard FIPS 146-2 that allows both **OSI** and **TCP/IP** protocol suites.

Positive Displacement Flowmeter a type of meter that measures volumes of fluid by counting repeated filling and discharging of known fixed volumes.

Positive Displacement Pump a reciprocating or rotary pump that produces a constant flow.

POSIX Portable Operating System Interface for uniX, a standard for the application programming interface to an operating system formalized as IEEE 1003 and ISO/IEC 9945.

POST (**Power On Self Test**) a diagnostic procedure that runs automatically when a device is turned on.

Post Office Protocol or **POP3**, an application-layer protocol used to retrieve e-mail from a remote server over a TCP/IP network.

POTS Plain Old Telephone Service, basic telephone service with limited features.

Power Factor a characteristic of alternating current electric systems expressed as a dimensionless number between 0 and 1 representing the ratio of real power to apparent power. Resistive loads like heaters have a power factor of 1; inductive and capacitive loads like ballasts, motors, and computers, have power factors that are less than 1 and require higher currents to transfer a given quantity of power.

Power Line Carrier or **Power Line Communications,** technology for transmitting data over the same wires used for electrical power.

Power On Self Test a diagnostic procedure that runs automatically when a device is turned on.

Power over Ethernet technology that allows devices to receive 48 volt DC electrical power via Ethernet cables in accordance with standard IEEE 802.3AF; this has the advantage of saving on installation effort and complexity by only requiring one LOW VOLTAGE cable to the device, saving energy by eliminating some power supplies, and can be used to deliver power with UNINTERRUPTIBLE POWER SUPPLY backup. See also POWER SOURCING EQUIPMENT and POWERED DEVICE.

Power Quality Monitoring systems to measure electrical power parameters like voltage, frequency, transients, etc.

Power-Shades devices that control the amount of light entering an area by raising or lowering a rolled, pleated, or cellular shade under automatic or manual control. See also AUTOMATED BLINDS, ROLLER-SHADE.

Power Sourcing Equipment (PSE), a device that selectively provides electrical power in a POWER OVER ETHERNET environment. An endspan PSE provides power as part of as an additional function in an Ethernet SWITCH; a midspan PSE is typically a standalone device or a patch panel that inserts power and can work connections from multiple switches.

Powered Device (PD), a devices that utilized power provided by POWER OVER ETHERNET.

Powerline see POWER LINE CARRIER.

PP (Protection Profile), a COMMON CRITERIA specification document that identifies security requirements relevant to a given group of users for a particular purpose and includes security objectives, security related functional requirements, information assurance requirements, assumptions, and rationale. Each profile defines covers a class of security device such as Anti-Virus, **Biometric** measures, **Certificate Management**, etc.

PPM Private Placement Memorandum or **Project and Program Management**.

PPP Point-to-Point Protocol, a data link layer protocol for establishing a direct connection between two nodes with features for authentication, encryption, and compression; typically used for dial-up access to the Internet or other TCP/IP networks. See RFC 1661 for details.

PPPoE or **PPP over Ethernet**, a network protocol for encapsulating **PPP** frames within **Ethernet** frames and establishing **IP** connections using the authentication, encryption and compression features of PPP. See RFC 2516 for details.

PQM (Power Quality Monitoring) systems to measure electrical power parameters like voltage, frequency, transients, etc.

Predictive Dialer a system or **Automatic Call Distributor** feature automatically dials outgoing telephone calls and connects them to customer service agents as they are answered. The system 'predicts' the likelihood that a call will be answered and when an agent will be available in order to maximize agent utilization.

Premises Equipment equipment installed in a customer's location including security systems, fire alarm control panels and access control systems.

Presence Information status indicating the ability and willingness of a user to communicate via **instant messaging**, telephone or other media. Common status values include 'free for chat,' 'busy,' 'away,' 'do not disturb,' and 'out to lunch.'

Private Branch Exchange a telephone switch (or exchange) owned by a private business.

Private Operating Mode fire alarm signaling that is only audible or visible to persons directly concerned with the implementation and direction of emergency actions and procedures. Contrast to **Public Operating Mode**.

Proactive Maintenance a maintenance strategy for maximizing the reliability of equipment by corrective actions aimed at failure root causes, not failure symptoms, faults, machine wear, or fixed schedules. Proactive maintenance involves setting quantifiable standards relating to each root cause, maintenance programs to control the root cause properties, and measurement or monitoring of each root cause property.

Production System a system that is used to process real transactions as contrasted to systems that are used for development or testing.

Productivity the amount of output created per unit of resource used. Labor productivity is output per worker or output per labor-hour; total factor productivity includes both labor and capital.

Profibus Process **Fieldbus**, a communication bus standard used in factory automation.

Profitable-to-Promise software that evaluates the profitability of individual orders based on supply chain management, customer relationship management and financial data.

Program 1) a *document* summarizing the results of a needs analysis including requirements and solutions for building space and occupancy; 2) a standalone unit of *software*; 3) the *process* of configuring or commanding a device; 4) a group of related *projects*.

Programmable Logic Controller a small computer used for automation of industrial processes or other control applications.

Progressive Scanning a video display technique where all lines are refreshed in each pass; contrast to **interlace**.

Project the process of planning, design, documentation, contracting, construction and management that results in a facility.

Project and Program Management business processes and systems for managing individual projects and programs or multiple related projects.

Property Management System an application system used to manage real-estate assets, leases, tenants, maintenance, and related business transactions for a property owner or developer. In the hospitality industry this term is used to describe more detailed oriented systems that support the day-to-day operation of a specific property for functions like guest registration, guest request tracking, etc.

Proportional-Integral-Derivative a type of digital or analog controller or control algorithm that uses a signal, its cumulative value, and rate of change in feedback calculations to adjust the controlled quantity to a given SETPOINT value.

Protection Profile (PP), a COMMON CRITERIA specification document that identifies security requirements relevant to a given group of users for a particular purpose and includes security objectives, security related functional requirements, information assurance requirements, assumptions, and rationale. Each profile defines covers a class of security device such as Anti-Virus, BIOMETRIC measures, CERTIFICATE MANAGEMENT, etc.

Protocol an agreement that enables communications. Protocols between humans include greeting styles like kiss, bow, or shake hands. Protocols between computers and other devices are frequently organized in layers or stacks, see OSI MODEL.

Protocol Implementation Conformance Statement a document that identifies the specific protocol options implemented by a device.

PRT (**Platinum Resistance Thermometer**) same as RESISTANCE TEMPERATURE DETECTOR.

PS/2 or **PS-2**, originally IBM Personal System/2 computers, currently used as shorthand for the mini-DIN-6 keyboard and mouse connector standard.

PS2 Sony Play Station 2 video game console.

PSAP (Public Safety Answering Point) a municipal or county emergency communication center that directs 9-1-1 or other emergency calls to appropriate police, fire, and medical services agencies.

PSE (Power Sourcing Equipment), a device that selectively provides electrical power in a POWER OVER ETHERNET environment. An endspan PSE provides power as part of as an additional function in an Ethernet SWITCH; a midspan PSE is typically a standalone device or a patch panel that inserts power and can work connections from multiple switches.

PSI Pounds per Square Inch.

PSM Process Safety Management.

PSN Packet Switched Network, today this means an **IP** network where chunks of data are dynamically switched or routed over shared links between end-points. Contrast to CIRCUIT SWITCHING.

PSTN the Public Switched Telephone Network.

PSU Power Supply Unit.

Psychometrics a field of study dealing with theory and techniques of educational and psychological measurement.

Psychrometer a measuring device that includes a dry-bulb and a wet-bulb thermometer.

Psychrometrics study of the physical and thermodynamic properties of gas-vapor mixtures. Psychrometric charts or software are used to relate the relative humidity, DEW POINT, and ENTHALPY to a combination of wet bulb and dry bulb temperature readings.

PTP Point-to-Point or **Profitable-to-Promise**.

PTT Post Telephone Telegraph or Push To Talk.

PTZ Pan, Tilt, Zoom, a type of video camera that can be controlled remotely.

Public Key Infrastructure a system for trusted third party vetting of, and vouching for, user identities and associating public encryption keys to users based on **X.509** or other standards. Public keys are typically in a **Digital Certificate** that contains a **Digital Signature** from a **Certificate Authority**.

Public Operating Mode audible or visible signaling to occupants or inhabitants of the area protected by the fire alarm system. Contrast to **Private Operating Mode**.

Public Safety Answering Point a municipal or county emergency communication center that directs 9-1-1 or other emergency calls to appropriate police, fire, and medical services agencies.

Public Utilities Commission a government regulatory agency responsible for the oversight of utilities at the state or local level.

PUC (Public Utilities Commission) a government regulatory agency responsible for the oversight of utilities at the state or local level.

Puck slang for a location-tracking radio-transmitter device.

Pull Station a device for manually signaling a fire alarm.

Pulse-Code Modulation a digital representation of an analog signal where the magnitude of the signal is sampled regularly at uniform intervals.

Push the process of sending data from a server to a client automatically; contrast to the more common **pull** operation based on client requests.

PV (Photovoltaic) or **Solar Cells,** materials that convert light into electrical power. See **SOLAR PANEL.**

PVC **PERMANENT VIRTUAL CIRCUIT** or **POLY VINYL CHLORIDE.**

PVD Physical Vapor Deposition, a technique used to deposit a very thin film of a solid or liquid onto a surface, used to create semiconductors and in other types of manufacturing.

PVR Personal Video Recorder.

PWAC Present Worth of Annual Charges, current value of a stream of future payments.

PWM Pulse Width Modulation.

Q

QC Quality Control.

QoE (Quality of (user) Experience) a subjective measure of a customer's experiences with a vendor or system.

QoS (Quality of Service) control mechanisms that can provide different priority to different users, or guarantee a certain performance level for a data flow.

QR Quick Response.

QSA Qualified Security Assessor, a firm certified to conduct data security assessments by the PCI Security Standards Council. For more information see www.pcisecuritystandards.org.

Quad Band a mobile phone or other device that is capable of **GSM** signaling on each of the four different radio frequencies used in different parts of the world.

Quality of (user) Experience a subjective measure of a customer's experiences with a vendor or system.

Quality of Service control mechanisms that can provide different priority to different users, or guarantee a certain performance level for a data flow.

Quicktime or **Apple Quicktime**, a multimedia framework, formats, and associated software developed by Apple. Quicktime format has become part of the MPEG-4 standard and the Quicktime player software supports a variety of formats.

R

R&A facility Repair and Alteration.

R&D Research & Development.

R-number chemical refrigerants are assigned an R number based on their molecular structure by **ASHRAE**. See Fig. 3 on page 280 for examples.

RA Return Air.

Raceway a type of conduit designed to support and protect wiring or fiber optic cables.

Radiance a measure of the amount of light emitted or reflected from a particular area that will be received by an optical system looking at the surface from some angle of view.

Radiant Heating any technology where energy emitted (radiating) from a heat source warms people and other objects directly rather than heating the air via convection. High temperature radiant heaters are also known as Infrared Heaters and typically used in open environments; low temperature radiant heating takes many forms including **UNDERFLOOR HEATING** and **CHILLED BEAMS** which can also be used for heating.

Radio Frequency electromagnetic spectrum in a variety of frequency bands including microwave and higher frequencies. See Fig. 8 on page 284.

Radio Frequency Identification an automatic identification technology that uses chips or tags that can be attached to, or incorporated into, a product, animal, or person and read remotely via radio. See also ACTIVE RFID and PASSIVE RFID.

Radio Frequency Interference electromagnetic interference by radio signals disrupting the operation of other electronic equipment.

Radiology Information System application system for storing X-ray and other medical images, and associated information.

RADIUS Remote Authentication Dial-In User Service, the RFC-2865 protocol for AUTHENTICATION, AUTHORIZATION, AND ACCOUNTING that is used for network access, **SIP,** etc. or a device that implements these services.

RAG Return Air Grate.

RAID or **Reliable Arrays of Inexpensive Disks,** a technology that combines multiple disk drives to improve capacity, reliability, availability, or performance. For example, RAID 0 uses disk striping to improve performance, RAID 1 uses mirroring to improve reliability.

RAM RANDOM ACCESS MEMORY or RISK ASSESSMENT METHODOLOGY.

RAMCAP or **Risk Analysis and Management for Critical Assets Protection,** a specific RISK ASSESSMENT METHODOLOGY developed by the ASME for the Department of Homeland Security.

Random Access Memory a form of electronic memory that may be accessed and changed quickly. Semiconductor random access memory is volatile and does not retain its contents if power is lost; contrast to READ ONLY MEMORY or FLASH.

Rapid Spanning Tree Protocol an evolution of the Spanning Tree Protocol that enables faster routing table updates following a change in network TOPOLOGY.

RAT Return Air Temperature.

RAW a digital image file format that is not compressed. Raw mode images are higher quality but require more storage than images in compressed formats like **JPEG**.

RCM Reliability Centered Maintenance, a risk management based process for preventative maintenance standardized in SAE JA1011. For more information see www.sae.org.

RCx or **Retro-Commissioning**, commissioning of existing buildings, see COMMISSIONING.

RDB (Relational Database Management System) a system for storing large amounts of data organized into tables containing rows and columns that may be linked together by shared data values representing the relationships between entries.

RDBMS (Relational Database Management System) a system for storing large amounts of data organized into tables containing rows and columns that may be linked together by shared data values representing the relationships between entries.

Read Only Memory electronic memory that may not be changed and retains its contents even with the power off.

Real-Estate Investment Trust a type of specialized investment entity that reduces or eliminates corporate taxes in return for being required to distribute 90% of income to its investors.

Real Estate Transaction Standard (RETS), a series of **XML** based standards designed to facilitate data transfer between **MULTIPLE LISTING SERVICE**s and other end points as well as facilitating access to listing information. For more information see www.rets.org.

Real Property Unique Identifier (RPUID), an **OSCRE** standard for an identification code that can be assigned to properties, buildings, structures and linear assets such as roadways, runways, tunnels, parking lots, etc.

Real-Time 1) systems designed to comply with specific time constraints; 2) simultaneous or nearly simultaneous activity.

Real-time Operating System any **OS** that includes special features for handling time critical tasks.

Real Time Streaming Protocol a protocol for streaming audio and video data that was developed by the **IETF** and published as RFC 2326.

Reduced Function Device a simple or single function device in a building automation system like a temperature sensor or occupancy sensor. Contrast to a FULL FUNCTION DEVICE.

Reduced Instruction Set Computing a type of processor designed to support a smaller number of different instructions but to execute each instruction more quickly, contrast to **CISC**.

Refrigerant a chemical compound used in a heat cycle where it undergoes a phase change from a gas to a liquid and back to gas; the refrigerant absorbs heat by evaporating at a low pressure and it gives up heat by condensing at a higher pressure. Each refrigerant is assigned an R-NUMBER for identification, see Fig. 3 on page 280 for examples.

Regenerative Elevators elevators equipped with power harvesting equipment to capture the mechanical energy from braking and convert it into electrical energy.

Registry a specialized type of database used to store configuration information within a system or OPERATING SYSTEM. For example, the Windows Registry.

REIT (**Real-Estate Investment Trust**) a type of specialized investment entity that reduces or eliminates corporate taxes in return for being required to distribute 90% of income to its investors.

Related Key Attack any form of cryptographic code breaking activity where the attacker can exploit a mathematical relationship between key values in order to recover the encryption key. For example, WIRED EQUIVALENT PRIVACY is vulnerable to related-key attacks.

Relational Control a control technology for energy use optimization including demand based controls based on the EQUAL MARGINAL PERFORMANCE PRINCIPLE and intelligent iterative controls based on iterative problem solving techniques. Contrast to the PROPORTIONAL INTEGRAL DERIVATIVE controls traditionally used in HVAC.

Relational Database Management System a system for storing large amounts of data organized into tables containing rows and columns that may be linked together by shared data values representing the relationships between entries.

Reliability the probability that a system will perform its intended function during a specified period of time under stated conditions; in transaction processing, reliability is evaluated in terms of the likelihood of losing or duplicating a transaction.

Remote Function Call (RFC), a distributed programming technique where a program calls a software module or function that is actually running on another system such as SERVER.

Remote Procedure Call a programming technique for procedure calls between software programs running on different computers connected via a network.

Replication a process for making or maintaining multiple copies of a database or other information. Typically replicated databases are used to increase RELIABILITY; replicated data at multiple locations may be used for disaster backup.

Request For Comment (RFC), 1) a type of document used by many organizations to solicit input on new policies, systems, or regulations; 2) the IETF assigns RFC numbers to documents during the standard development process and standards continue to be known by the RFC number after they are approved.

Request For Information a procurement document asking for specific information, may be followed by a **RFP**, or REQUEST FOR QUOTATION, or an addendum to an existing contract.

Request For Proposal (RFP), a procurement document requesting a specific proposal. In some cases a proposal will be followed by vendor selection and a contract, in other cases the proposal may be followed by a REQUEST FOR QUOTATION.

Request for Qualifications (RFQ), a procurement document intended to make potential suppliers aware of an opportunity and solicit information on qualifications (background, past performance, financial condition, resources, etc.) and availability prior to issuing a REQUEST FOR PROPOSAL.

Request for Quotation (RFQ), a procurement document requesting a specific quotation typically based on a previous proposal.

Resistance Temperature Detector a sensor that uses the change in electrical resistance of a material like platinum with changing temperatures as a way to measure temperature.

Resolution 1) a *measure* of the amount of detail in an image or display in terms of dots per inch, pixels, or scan lines; 2) the *process* of retrieving information as in domain name resolution; 3) the *process* of solving a problem or reaching an agreement.

Resource a general term for the equipment, services, or personnel required for a task.

REST REpresentational State Transfer, any simple web-based interface that uses **XML** and **HTTP** without the complexity of the message exchange patterns used in **SOAP**.

Retrofit the process of adding new technologies or features to older systems or upgrading a building to meet additional requirements. For example, a seismic retrofit for improved earthquake safety.

RETS (Real Estate Transaction Standard), a series of **XML** based standards designed to facilitate data transfer between **MULTIPLE LISTING SERVICE**s and other end points as well as facilitating access to listing information. For more information see www.rets.org.

Return Air air that is drawn out of the building by the HVAC system; return air may be exhausted from the building or recirculated within the building.

RF RADIO FREQUENCY or Return Fan, depending on the context.

RfC Reference Concentration, the concentration of a chemical in air that is very unlikely to have adverse effects if inhaled continuously over a lifetime.

RFC REMOTE FUNCTION CALL or **REQUEST FOR COMMENT**.

RfD Reference Dose, estimated daily exposure to a toxic non-carcinogen that is not likely to cause harmful effects during a lifetime.

RFD (Reduced Function Device) a simple or single function device in a building automation system like a temperature sensor or occupancy sensor. Contrast to a **FULL FUNCTION DEVICE**.

RFDC Radio Frequency Data Communications or Radio Frequency Data Collection.

RFI RADIO FREQUENCY INTERFERENCE or **REQUEST FOR INFORMATION**.

RFID (Radio Frequency Identification) an automatic identification technology that uses chips or tags that can be attached to, or incorporated into, a product, animal, or person and read remotely via radio. See also ACTIVE RFID and PASSIVE RFID.

RFP (Request For Proposal), a procurement document requesting a specific proposal. In some cases a proposal will be followed by vendor selection and a contract, in other cases the proposal may be followed by a REQUEST FOR QUOTATION.

RFQ REQUEST FOR QUALIFICATIONS or REQUEST FOR QUOTATION.

RG6 coaxial cable used for cable TV and other applications.

RGB Red Green Blue, the color components used in video signaling or imaging.

RH Relative Humidity.

Rich Text Format (RTF), a Microsoft proprietary file format used to exchange formatted word processing documents between systems. Most word processors can read and write rich text format and converters are available for HTML and other formats.

Rijndael the original name for block cipher now known as **AES**. Rijndael comes from a combination of the names of the inventors Rijmen and Daemen.

RIS (Radiology Information System) application system for storing X-ray and other medical images, and associated information.

RISC (Reduced Instruction Set Computing) a type of processor designed to support a smaller number of different instructions but to execute each instruction more quickly, contrast to **CISC**.

Risk 1) the potential impact, loss, or exposure associated with a threat, risk can be quantified based on the probability or likelihood of the threat, the probability of vulnerability to the threat, and the potential impact or loss; 2) the probability of an anticipated loss.

Risk Assessment Methodology a process for evaluating risks and risk mitigation strategies. There are many methods and tools available including the RAM series from Sandia Labs.

RJ-11 four position modular connector used for telephones and similar devices.

RJ-45 eight position modular connector used for multi-line telephone and ETHERNET cabling systems.

RLA Rated Load Amps.

RLL Run Length Limited, a type of encoding used on computer disks to minimize errors.

ROA Return on Assets.

ROE Return on Equity.

ROI Return On Investment.

Roller Shade an automatic device that consists of fabric wrapped around a horizontal tube that contains a tubular motor. The motor rotates to change the shade height to control daylight, solar heat gains, view, glare, privacy, or building appearance.

ROM (Read Only Memory) electronic memory that may not be changed and retains its contents even with the power off.

Room Controller a controller for a specific room, usually for lighting.

Root Cause the process of root cause analysis involves researching the events leading up to a problem or outage and identifying the original source of the problem.

Router a device that connects sub-networks together to form a larger network by selectively forwarding packets between sub-networks. In an IP network, a router may perform **NETWORK ADDRESS TRANSLATION** as part of this process.

RPC (Remote Procedure Call) a programming technique for procedure calls between software programs running on different computers connected via a network.

RPM Revolutions Per Minute.

RPP Remote Power Panel.

RPUID (Real Property Unique Identifier), an **OSCRE** standard for an identification code that can be assigned to properties, buildings, structures and linear assets such as roadways, runways, tunnels, parking lots, etc.

RQE Request to Exit.

RS-232 EIA standard RS-232-C or ITU V.24, serial data communications interface, see **TIA-232**.

RS-485 or EIA-485, a physical interface specification for a two-wire, half-duplex, multipoint serial connection.

RSF Rentable Square Feet.

RSS a family of XML based formats for sharing or syndicating Web content. Includes Really Simple Syndication, Rich Site Summary, and RDF Site Summary.

RSTP (Rapid Spanning Tree Protocol) an evolution of the Spanning Tree Protocol that enables faster routing table updates following a change in network TOPOLOGY.

RT Radiographic, Real-time, or Return Temperature.

RTD (Resistance Temperature Detector) a sensor that uses the change in electrical resistance of a material like platinum with changing temperatures as a way to measure temperature.

RTE Request to Exit, a device located inside a secure area that senses a person approaching a secured door to allow exit without generating an alarm.

RTF (Rich Text Format), a Microsoft proprietary file format used to exchange formatted word processing documents between systems. Most word processors can read and write rich text format and converters are available for HTML and other formats.

RTFM Read The Fine Manual.

RTN Return to Normal.

RTO Recovery Time Objective, a measure used in a BUSINESS CONTINUITY PLAN for the allowable time required to restore a business process or system following a disruption in order to keep the organization afloat.

RTOS (Real-time Operating System) any OS that includes special features for handling time critical tasks.

RTP Real-Time Pricing.

RTSP (Real Time Streaming Protocol) a protocol for streaming audio and video data that was developed by the IETF and published as RFC 2326.

RTT Radio Transmission Technology.

RTU Remote Terminal Unit.

RTX Real-time Extension.

S

S/PDIF (Sony/Philips Digital Interface Format) hardware and low-level protocol specifications for connecting digital audio devices standardized as IEC 958 type II, part of IEC-60958.

S/W Software.

S-Video a video interface standard with separate color (chrominance) and intensity (luminance) signals. S-Video offers better signal quality than COMPOSITE VIDEO.

SA SMART ACTUATOR or System Air.

SAB SOUND ATTENUATION BATT, Sound Attenuation Blanket, or Sound Attenuation Board.

SAC Security and Access Control, combined system for SECURITY SYSTEM and ACCESS CONTROL.

SACD (Super Audio Compact Disk) a read-only optical disc designed to provide much higher fidelity digital audio than the original COMPACT DISC.

SAD (Seasonal Affective Disorder) an affective, or mood disorder, with depressive symptoms linked to shorter days and longer nights during the winter in the northern hemisphere or the summer in the southern hemisphere.

SAFB (Sound Attenuation Fire Batt), a type of insulation with acoustical, thermal, and fire resistance properties.

Safe-Room or **Shelter-in-Place Room,** spaces within a building that are designed to provide emergency protection. Depending on the expected threats, they may have limited outdoor air exchange and limited air exchange with the rest of the building.

Safety Instrumented System an industrial process equipped with an independent system of sensors, controllers, and control elements designed to return the system to a safe state when certain pre-determined conditions are violated.

Safety Integrity Level IEC 61508 defines four levels of safety performance for a safety function from SIL1 to SIL4 and details the requirements necessary to achieve each safety integrity level.

Sales Force Automation application systems used to track customers and prospects in support of direct marketing and sales processes.

SAMA Scientific Apparatus Makers Association.

SAME (Specific Area Message Encoding), an alert and notification message protocol used by the Emergency Alert System (EAS) in the US and Canada. Each message includes the originator of the alert, the event type, the areas affected, the duration, and the time the alert was issued.

SAN (Storage Area Network) a technology for attaching computers and storage devices or storage systems that communicates over a network using block-based protocols. Similar to **NAS.**

SAP Security Access and Parking.

Sarbanes Oxley Act of 2002, US law requiring additional reporting, controls and procedures for public companies.

Sarbox slang term for Sarbanes-Oxley.

SAT Supply Air Temperature.

SAW (Surface Acoustic Wave) piezoelectric devices used in electronics circuits as filters, oscillators, transformers, or sensors based on the transduction of acoustic waves.

SBS (Sick Building Syndrome) a situation where a substantial percentage (i.e. over 20%) of the occupants of a building, or an area within a building, are affected by a set of symptoms that persist for at least two weeks, the symptoms dissipate soon after exiting the building, and the cause or causes are unknown.

SC (Substantial Completion) a milestone in the construction process that represents the point at which the client may request an occupancy permit in preparation for using the building and the start of any contractor warranties.

SCADA Supervisory Control and Data Acquisition, distributed measurement and control system technology linking Remote Terminal Units, Programmable Logic Controllers, and Human Machine Interface devices over a communication network. Widely used in process control and utility applications, rarely seen in building automation.

Scalable Vector Graphics (SVG), an open standard for describing two-dimensional vector graphics using XML. See www.w3.org/Graphics/SVG/ for more information.

SCAQMD (South Coast Air Quality Management District) the smog control agency for Los Angeles, Orange, Riverside and San Bernardino counties in California. For more information see www.aqmd.gov.

SCE Supply Chain Execution.

Scene Control pre-defined lighting levels for different uses of a space.

Schema a data model describing the information to be stored in a DATA-BASE for a specific problem domain in terms of how the data will be organized or structured and how data may be accessed or manipulated.

SCM Supply Chain Management.

SCMS (Smart Card Management System), a computer application for managing SMART CARDS, card applications, and related databases. The SCMS provides or links to applications for card issuance, KEY MANAGEMENT SYSTEM, card PERSONALIZATION, CUSTOMER RELATIONSHIP MANAGEMENT, etc.

SCP Supply Chain Planning.

SCPM Supply Chain Process Management.

SCS (Structured Cabling System), a standard hierarchical cable system infrastructure supporting multiple applications including voice, data, video, and BUILDING AUTOMATION SYSTEM; this has the advantage of reducing cabling costs and space requirements and facilitating integration. See standards **TIA-568** for commercial, **TIA-862** for building automation, **TIA-942** for data centers, and Fig. 11 on page 287.

SCSI (Small Computer System Interface) a standard physical interface and commands for data storage, commonly used for disk and tape drives.

SDACT (Serial Digital Alarm Communicator Transmitter) a type of DIGITAL ALARM COMMUNICATOR TRANSMITTER designed to accept serial digital inputs such as the printer output of a FIRE ALARM CONTROL PANEL.

SDH Synchronous Digital Hierarchy, optical networking standard for telephone and data traffic based on 155.52 Mbit/s signals and standardizes as INTERNATIONAL TELECOMMUNICATION UNION G.707 and G.708. The North American counterpart is **SONET**.

SDK Software Development Kit, a set of development tools and documentation that enables a software engineer to create applications that for a certain software package, framework, platform, computer system, etc.

SDLC (Synchronous Data Link Control) a communication protocol for layer two of the **OSI MODEL** developed by IBM; later became the basis for **HDLC**.

SDN Specially Designated National, individuals and companies whose assets are blocked by **OFAC**.

SDO (Standards Development Organization) any entity whose primary activities include developing, coordinating, or maintaining standards that address the interests of a wide base of users outside the organization.

SDSL (Symmetric Digital Subscriber Line) a form of **DSL** offering the same transmission speeds in both directions.

SDV (Switched Digital Video), techniques for efficiently distributing digital video via a network infrastructure with limited capacity. In HYBRID FIBER COAX networks all channels are delivered to the fiber optic node which dynamically assigns channels on the coax portion of the network based on user requests, thus makes more efficient use of the limit number of coax channels by only transmitting channels that are being watched. **IPTV** and other **IP** based video services may use UNICAST or MULTICAST STREAMING MEDIA protocols such as **RTSP**.

Search Engine Optimization the art of optimizing a website design and content in order to maximize the search result ranking for selected search terms.

Seasonal Affective Disorder an affective, or mood disorder, with depressive symptoms linked to shorter days and longer nights during the winter in the northern hemisphere or the summer in the southern hemisphere.

Seasonal Energy Efficiency Ratio a rating of air conditioner energy efficiency.

SECAM or **Séquentiel Couleur à Mémoire**, French for sequential color with memory, is an analog color television system first used in France.

Second Life an online virtual world and multiplayer game operated by Linden Labs.

Secure Hash Algorithm actually a family of related cryptographic hash functions designed by the National Security Agency and published as federal standards. SHA-1 is used in **TLS, SSL, PGP, SSH**, S/MIME, and **IPsec**; SHA-2 variants include SHA-224, SHA-256, SHA-384, and SHA-512.

Secure Shell an **Internet** protocol (RFC-4251) for establishing a secure channel between a client and a server computer that uses public-key cryptography to authenticate the computers, and optionally to authenticate the user, to protect the privacy of the data exchanged, and to protect the integrity of the data.

Secure Socket Layer a protocol for encrypting network communications that has been superseded by **Transport Layer Security**.

Security the condition of being protected against danger or loss with an emphasis on protection from dangers that originate from outside.

Security Certificate or **Public Key Certificate**, an electronic document that incorporates a digital signature used to bind together a public key and an identity. When used as part of a **public key infrastructure** system, the digital signature will be from a **Certificate Authority**.

Security Industry Association organization of security product manufacturers, see www.siaonline.org for more information.

Security Portal a **Portal** designed to give security personnel access to the information they need to assure the security and safety of the facility including status of fire and security devices, access controls, occupancy tracking, elevator status, video surveillance and recorders, etc. In some cases security may also provide off-hours monitoring of certain building automation system alarms.

Security System an application that facilitates monitoring and documenting security activities including the status of burglar alarms, intrusion alarms, perimeter security, fire alarms, closed circuit TV and video surveillance, access control systems, etc.

SEER (Seasonal Energy Efficiency Ratio) a rating of air conditioner energy efficiency.

Seismic Sensor a sensor that detects vibrations generated by a person walking or digging in its area of sensitivity.

Self-Healing a network or system that incorporates redundancy and the ability to continue operating even if some components or facilities fail.

Self-Test a diagnostic procedure that is run automatically.

SEO (Search Engine Optimization) the art of optimizing a website design and content in order to maximize the search result ranking for selected search terms.

Sequence of Operations the specification or operational definition for control system actions including a description of the system configuration, state definitions, procedures for system startup and shutdown, scheduled operations, and system response to different conditions.

Sequential Function Chart a graphical programming language used for programmable logic controllers standardized as IEC 848.

Sercos or **Serial Real-time Communication System,** IEC 61491 standard digital interface between controllers and drives in numerically controlled machines.

Serial Digital Alarm Communicator Transmitter a type of DIGITAL ALARM COMMUNICATOR TRANSMITTER designed to accept serial digital inputs such as the printer output of a FIRE ALARM CONTROL PANEL.

Serial Line Internet Protocol a way to encapsulate IP over serial lines, replaced by PPP.

Server a device on a network that stores data and provides information to other devices; a server may be a function of a device that has another purpose (an embedded server), a PC, a specialized computer, or a mainframe. Contrast to CLIENT.

Server Room a room designed to hold SERVERs and communications equipment. Typically a small room equipped with extra power and cooling capacity.

Server Virtualization technology for running multiple 'virtual' server images within a single larger computer.

Service Level Agreement a contract between two groups specifying the performance terms for services one group will provide to the other. May specify objectives for system AVAILABILITY and RELIABILITY.

Service Level Objective a target or objective for supply chain performance.

Service Oriented Architecture a design for loosely coupled network applications typically using WEB SERVICES. The SOA reference model is being standardized by Oasis, see www.oasis-open.org for more information.

Session Initiation Protocol a protocol developed by the **IETF** for managing multimedia user sessions, widely used for **VOIP** signaling.

Set Top Box end user control device for cable TV and other forms of interactive television.

Setpoint the target value for an automated control system.

SF Specialty Fabricator, Square Feet, or Supply Fan.

SFA (Sales Force Automation) application systems used to track customers and prospects in support of direct marketing and sales processes.

SFC (Sequential Function Chart) a graphical programming language used for programmable logic controllers standardized as IEC 848.

SFPE (Society of Fire Protection Engineers) a professional society for those practicing the field of fire protection engineering. For more information see www.sfpe.org.

SGML (Standard Generalized Markup Language) or ISO 8879:1986, is a meta-language for defining markup languages, SGML is used for text and database publishing in large organizations. HTML and XML are derived from SGML.

SHA (Secure Hash Algorithm) actually a family of related cryptographic hash functions designed by the National Security Agency and published as federal standards. SHA-1 is used in **TLS, SSL, PGP, SSH, S/MIME**, and **IPsec**; SHA-2 variants include SHA-224, SHA-256, SHA-384, and SHA-512.

Shade Control 1) a *system* for controlling the amount of external light entering a space by automatically or manually adjusting window shades or similar devices, see **Power-Shades, Roller Shade, Sun-Blinds**; 2) *regulation* to prohibit shading of **Solar Panel** systems by neighboring trees and shrubs, for example the California Solar Shade Control Act.

Shared Memory a computer configuration where part of the computer memory is dedicated to other functions such as video display.

Shared Tenant Services telecommunications and other services offered to tenants by a building operator.

Sharepoint a family of Microsoft collaboration and document management products.

Sheet Metal and Air Conditioning Contractors' National Association an international trade association and standards organization. For more information see www.smacna.org.

SHGC (Solar Heat Gain Coefficient), the fraction of solar radiation admitted through a window, door, or skylight and subsequently released as heat inside. Low SHGC glazing transmits less solar heat and reduces cooling loads; high SHGC glazing is more effective at collecting solar heat.

Shielded Twisted Pair a type of cabling with one or more twisted pairs of wire enclosed in a metallic foil shield that is grounded and designed to protect the wires from electromagnetic interference.

SHM (Structural Health Monitoring) systems for monitoring building structure.

Shop Drawing a detail drawing provided by the manufacturer or contractor to show how an item will be fabricated and/or installed as part of a construction project.

Short-cycling a condition in which a compressor or furnace is restarted immediately after it was turned off.

Short Message Service a form of text messaging on cellular telephones.

Shrinkage the retail term for the reduction in inventory caused by shoplifting or spoilage.

SI System Integrator/Systems Integration or **Système International**.

SIA Security Industry Association or **SIA Format**.

SIA Format a standard for alarm system communications interfaces developed by the **Security Industry Association** and formalized through **ANSI**; current version is SIA DC-03-1990.01 (R2003.10).

SIC or **Standard Industrial Classification,** codes for classifying business activities, being replaced by **NAICS**.

Sick Building Syndrome a situation where a substantial percentage (i.e. over 20%) of the occupants of a building, or an area within a building, are affected by a set of symptoms that persist for at least two weeks, the symptoms dissipate soon after exiting the building, and the cause or causes are unknown.

Sidelighting a Daylighting technique where natural light enters through a window. Sidelighting can be used in corner rooms and buildings with large perimeter areas but it provides illumination with strong directionality and may cause glare.

Signal to Noise Ratio the relative quantities of useful information or signal and other noise delivered by a communications channel.

SIL (Safety Integrity Level) IEC 61508 defines four levels of safety performance for a safety function from SIL1 to SIL4 and details the requirements necessary to achieve each safety integrity level.

Silo derogatory term for an application system or business unit with a narrow vertical focus on one specific function and limited ability to cooperate or integrate with other applications or business units.

Simple Mail Transfer Protocol standard protocol for e-mail transmission in TCP/IP networks.

Simple Network Management Protocol a protocol used by network management systems to monitor devices that are attached to a network. Part of the TCP/IP protocol suite, may be used by any type of device, not actually simple.

Simple Network Time Protocol a protocol for synchronizing the clocks of computer systems over a network.

Simple Object Access Protocol a protocol for exchanging XML-based messages between applications, normally using HTTP. See www.w3.org/TR/soap/ for more information.

Simple Service Discovery Protocol (SSDP), a protocol for Plug-and-Play device configuration over TCP/IP networks developed by the UPnP Forum.

Single Mode Fiber or **Single Mode Optical Fiber** (SMF), a type of optical fiber designed to carry only a single ray of light (mode) in the 1310 nm or 1550 nm transmission window that is capable of supporting high bandwidth applications and distances in the hundreds of kilometers. Singlemode fibers are typically used in long-distance telephony and cable TV applications while **MULTI-MODE FIBER** is used for most in-building applications.

Single Network Sign On same as **SINGLE SIGN ON**.

Single Sign On an access control system for multiple applications and systems that only requires a user to be authenticated once even through they are accessing multiple applications.

SIP (Session Initiation Protocol) a protocol developed by the **IETF** for managing multimedia user sessions, widely used for **VOIP** signaling.

SIS (Safety Instrumented System) an industrial process equipped with an independent system of sensors, controllers, and control elements designed to return the system to a safe state when certain pre-determined conditions are violated.

SKU (Stock Keeping Unit) an identifier used by merchants to permit the systematic tracking of products and services at the level of each variation, bundle, service, fee or attachment.

Skydome a **DAYLIGHTING** term meaning the dome of the sky, excluding the portion containing the solar disk.

SLA SERVICE LEVEL AGREEMENT, Software License Agreement, or **STEREOLITHOGRAPHY APPARATUS**.

SLC Single Loop Controller.

SLDC Single Loop Digital Control.

SLIP (Serial Line Internet Protocol) a way to encapsulate **IP** over serial lines, replaced by **PPP**.

SLO (Service Level Objective) a target or objective for supply chain performance.

SMACNA (Sheet Metal and Air Conditioning Contractors' National Association) an international trade association and standards organization. For more information see www.smacna.org.

Small Box Retailer a retail chain with a large number of small stores (under 5,000 square feet) including drug stores, gas stations, convenience stores, retail banks, small grocery stores, dry cleaners, etc.

Small Computer System Interface a standard physical interface and commands for data storage, commonly used for disk and tape drives.

Small Terminal Interoperability Platform a consortium that developed platform specifications for secure transaction devices based on smart cards and other technologies (see www.stip.org). In 2004 this group merged with the **GLOBAL PLATFORM DEVICE** consortium.

Smart Actuator an actuator that combines sensor and control capabilities in a single device.

Smart Building a building with advanced and integrated systems for building automation, life safety and telecommunications. See also **INTELLIGENT BUILDING**.

Smart Card a card or other device that contains a microprocessor and software to store information, carry out local processing on the data stored, and perform complex calculations in support of secure identification and a variety of other applications. Cards may be contact cards that must touch a card reader or contactless cards that use radio frequency signaling and may be read at a distance.

Smart Card Management System (SCMS), a computer application for managing **SMART CARDS**, card applications, and related databases. The SCMS provides or links to applications for card issuance, **KEY MANAGEMENT SYSTEM**, card **PERSONALIZATION**, **CUSTOMER RELATIONSHIP MANAGEMENT**, etc.

Smart Conference Room a conference room equipped with advanced multimedia technology and integrated environmental controls for lighting, temperature, and sound masking. May include advanced video conferencing systems with video cameras, video projectors, television sets, multiple microphones, and speakers.

Smart Escalator an escalator with advanced control systems features. For example, escalators that slow or stop when not in use.

Smart Glass glass or glazing that can be electrically controlled to vary the amount of light and heat transmitted or to change from transparent to opaque.

Smart Landscaping environment friendly practices for landscape design, plant selection, and care that conserve water and minimize chemical use and solid waste production. Landscaping may also be designed to minimize building energy requirements and to reduce fire hazards. See also XERISCAPING.

Smart Lighting energy efficient lighting using a variety of technologies including high efficiency lights, daylight harvesting, and controls that adjust lighting based on occupancy.

Smart Phone a handheld device that combines the functionality of a mobile phone and a PERSONAL DIGITAL ASSISTANT or other information appliance.

Smart Sensor a device that integrates sensors, control, and communications capabilities resulting in the ability to get better information with less bandwidth.

Smart Time Scheduling time based control of lighting and other environmental features in space that is not equipped with occupancy sensors.

Smart Window a window containing SMART GLASS.

SMARTcodes a project to automate code compliance checking for the I-Codes and Federal, state and locally adopted versions of those codes. For more information see www.iccsafe.org/SMARTcodes/.

SME Subject Matter Expert.

SMF (Single Mode Fiber) or **Single Mode Optical Fiber** (SMF), a type of optical fiber designed to carry only a single ray of light (mode) in the 1310 nm or 1550 nm transmission window that is capable of supporting high bandwidth applications and distances in the hundreds of kilometers. Singlemode fibers are typically used in long-distance telephony and cable TV applications while MULTI-MODE FIBER is used for most in-building applications.

SMIL (Synchronized Multimedia Integration Language) is a markup language for multimedia presentations based on XML that defines timing, layout, animations, visual transitions, and media embedding, etc. For more information see www.w3.org/AudioVideo/.

Smoke Control System the system that controls mechanical smoke control equipment including fans and SMOKE DAMPERs based on automatic alarm signals from the FIRE ALARM CONTROL PANEL or manual commands from the FIREFIGHTERS SMOKE CONTROL STATION.

Smoke Damper special UL LISTED dampers located in ducts at the perimeter of a SMOKE ZONE and powered by an emergency power source. Smoke dampers are subject to more stringent leakage tests than standard control dampers and typically have switches to indicate fully open or closed position.

Smoke Zone or **Smoke Control Zone**, 1) a *space* within a building where smoke may be controlled through compartmentalization and pressurization; 2) the *specific* HVAC zone in which smoke has been detected.

SMP (Symmetrical Multiprocessing) a type of computer system architecture with two or more identical processors (**CPUs**) and shared memory.

SMR (Specialized Mobile Radio), 1) *radio service*, a conventional two-way radio or trunked radio system licensed in the 800 or 900 MHz band; 2) *radio band* including 806-824, 851-869 MHz licensed by the FCC for Specialized Mobile Radio Service, iDEN, and Sprint/Nextel.

SMS (Short Message Service) a form of text messaging on cellular telephones.

SMT (Surface Mount Technology) a fabrication technique where components mount directly on printed circuit boards without inserting leads in holes or sockets.

SMTP (Simple Mail Transfer Protocol) standard protocol for e-mail transmission in TCP/IP networks.

SNMP (Simple Network Management Protocol) a protocol used by network management systems to monitor devices that are attached to a network. Part of the TCP/IP protocol suite, may be used by any type of device, not actually simple.

SNR (Signal to Noise Ratio) the relative quantities of useful information or signal and other noise delivered by a communications channel.

SNTP (Simple Network Time Protocol) a protocol for synchronizing the clocks of computer systems over a network.

SNVT (Standard Network Variable Type) a LonTalk protocol data element.

SOA (Service Oriented Architecture) a design for loosely coupled network applications typically using Web Services. The SOA reference model is being standardized by Oasis, see www.oasis-open.org for more information.

SOAP (Simple Object Access Protocol) a protocol for exchanging XML-based messages between applications, normally using HTTP. See www.w3.org/TR/soap/ for more information.

SOC Security Operations Center.

Society of Fire Protection Engineers a professional society for those
practicing the field of fire protection engineering. For more information
see www.sfpe.org.

SoftLogic PC-based Logic Control in software.

Softstart a control feature that provides the ability to start fans or other
devices slowly in order to minimize noise, vibration, and current draw.

SOHO Small Office/Home Office.

SoIP Storage over IP, see ɪSCSI and **FCIP**.

Solar Chimney a glass multi-story vertical shaft on the south façade of
a building used for heat extraction by cross ventilation from the cooler
north side of the building.

Solar Heat Gain Coefficient (SHGC), the fraction of solar radiation
admitted through a window, door, or skylight and subsequently released
as heat inside. Low SHGC glazing transmits less solar heat and reduces
cooling loads; high SHGC glazing is more effective at collecting solar
heat.

Solar Panel a flat solar energy collector that may be either a thermal
collector used to heat water or air or a **Photovoltaic** module used to
generate electricity.

Solid-State Disk or **Solid-State Drive**, a disk drive equivalent that uses
semiconductor memory technology instead of a magnetic disk for
faster access, lower power requirements, or smaller size. Semiconductor
memories use for these applications may be volatile **RAM** with a built in
battery or non-volatile **FLASH** memory.

Solid-State Lighting lighting that uses semiconductor materials to convert
electricity into light including **LIGHT-EMITTING DIODE** and **ORGANIC
LIGHT-EMITTING DIODE** technology.

SONET Synchronous Optical Networking, networking standard for telephone and data traffic based on 51.84 Mbit/s signals and standardized as GR-253-CORE from **Telcordia**. SONET is primarily North American; the equivalent for the rest of the world is **SDH**.

Sony/Philips Digital Interface Format hardware and low-level protocol specifications for connecting digital audio devices standardized as IEC 958 type II, part of IEC-60958.

SOO (Sequence Of Operations) the specification or operational definition for control system actions including a description of the system configuration, state definitions, procedures for system startup and shutdown, scheduled operations, and system response to different conditions.

SOP Standard Operation Procedure.

Sound Attenuation Batt or **Sound Batt,** a type of insulation designed for noise control.

Sound Attenuation Fire Batt (SAFB), a type of insulation with acoustical, thermal, and fire resistance properties.

Sound Card an add-on card, circuit, or feature for audio processing typically including the ability to convert audio signals between audio and digital.

Sound Masking technology for adding air conditioning-like sounds or white noise in a specific area to mask or cover-up human speech or to eliminate distractions of other sounds.

Sound Transmission Class (STC) a rating of how well a building partition attenuates airborne sound; STC is roughly equal to the decibel noise reduction provided by a partition.

South Coast Air Quality Management District the smog control agency for Los Angeles, Orange, Riverside and San Bernardino counties in California. For more information see www.aqmd.gov.

SOW Statement of Work.

SOX (Sarbanes-Oxley) Act of 2002, US law requiring additional reporting, controls and procedures for public companies.

SP Static Pressure.

SP50 the ISA (formerly Instrumentation Society of America) FIELDBUS standard or ANSI/ISA-S50 developed by Standards & Practice Committee 50.

Spanning Tree Protocol a network routing protocol used for a bridged LAN, see IEEE 802.1D and RAPID SPANNING TREE PROTOCOL.

SPC (Statistical Process Control) a method of quality control in manufacturing using statistical tools to monitor the process.

Specialized Mobile Radio (SMR), 1) *radio service*, a conventional two-way radio or trunked radio system licensed in the 800 or 900 MHz band; 2) *radio band* including 806-824, 851-869 MHz licensed by the FCC for Specialized Mobile Radio Service, iDEN, and Sprint/Nextel.

Specific Area Message Encoding (SAME), an alert and notification message protocol used by the Emergency Alert System (EAS) in the US and Canada. Each message includes the originator of the alert, the event type, the areas affected, the duration, and the time the alert was issued.

SPOC Single Point Of Contact.

SPOF Single Point of Failure.

Sprinkler Flow Switch a sensor in a fire sprinkler line that notifies the FIRE ALARM CONTROL PANEL of water flow in a sprinkler line.

Spyware software designed to intercept information or use the resources of a computer without the user's consent or knowledge.

SQC (Statistical Quality Control) a method of quality control in manufacturing using statistical tools to monitor the finished product.

SQL (Structured Query Language) a computer language for manipulating information stored in a relational database, also used as a synonym for relational database.

SS SMART SENSOR or Sustainable Sites.

SSD (Solid-State Disk) or **Solid-State Drive**, a disk drive equivalent that uses semiconductor memory technology instead of a magnetic disk for faster access, lower power requirements, or smaller size. Semiconductor memories use for these applications may be volatile **RAM** with a built in battery or non-volatile **FLASH** memory.

SSDP (Simple Service Discovery Protocol), a protocol for **PLUG-AND-PLAY** device configuration over **TCP/IP** networks developed by the **UPnP FORUM**.

SSH (Secure Shell) an **INTERNET** protocol (RFC-4251) for establishing a secure channel between a client and a server computer that uses public-key cryptography to authenticate the computers, and optionally to authenticate the user, to protect the privacy of the data exchanged, and to protect the integrity of the data.

SSL SECURE SOCKET LAYER or **SOLID-STATE LIGHTING**.

SSO (Single Sign On) an access control system for multiple applications and systems that only requires a user to be authenticated once even through they are accessing multiple applications.

SSP Shared Service Provider.

SSPC 135 ASHRAE Standing Standard Project Committee 135, the group that develops and maintains **BACNET** standards.

ST Supply Temperature.

Stack 1) *general* a pile of something; 2) *communications* a **Protocol** stack, see **OSI Model**; 3) *plumbing* a drain stack is a vertical combination of drains and vents.

Stack Effect vertical airflow between floors of a building caused by temperature differences between indoor and outdoor air. When indoor air is warmer than outdoor air it rises and escapes though the upper parts of the building shell as cool air enters the lower parts of the building; the flow is reversed if indoor air is cooler than outdoor air.

Stakeholder any person or organization with an interest in a given process or entity.

Standard for the Exchange of Product Model Data or STEP is the common name for ISO10303 Industrial automation systems and integration - Product data representation and exchange, a standard for exchanging data between **CAD** and **CAM** systems.

Standard Generalized Markup Language or ISO 8879:1986, is a metalanguage for defining markup languages, SGML is used for text and database publishing in large organizations. HTML and XML are derived from SGML.

Standard Network Variable Type a **LonTalk** protocol data element.

Standards Development Organization any entity whose primary activities include developing, coordinating, or maintaining standards that address the interests of a wide base of users outside the organization.

Star Network a network topology where connections branch out from central nodes like legs on a starfish.

Starved Box a variable air volume box where the actual flow is less than the desired flow, even though damper is 100% open.

Static Transfer Switch a device that can transfer an electrical load between AC power sources quickly and without power interruption. For example, switching a computer system between utility power and backup power.

Statistical Process Control a method of quality control in manufacturing using statistical tools to monitor the process.

Statistical Quality Control a method of quality control in manufacturing using statistical tools to monitor the finished product.

STB (Set Top Box) end user control device for cable TV and other forms of interactive television.

STC (Sound Transmission Class) a rating of how well a building partition attenuates airborne sound; STC is roughly equal to the decibel noise reduction provided by a partition.

STEP (STandard for the Exchange of Product model data) or STEP is the common name for ISO10303 Industrial automation systems and integration - Product data representation and exchange, a standard for exchanging data between **CAD** and **CAM** systems.

Stereolithography Apparatus a device that performs stereolithography, a technology used for rapid manufacturing and rapid prototyping to produce parts with high accuracy and good surface finish automatically from **3D CAD** models.

STIP (Small Terminal Interoperability Platform) a consortium that developed platform specifications for secure transaction devices based on smart cards and other technologies (see www.stip.org). In 2004 this group merged with the **GLOBAL PLATFORM DEVICE** consortium.

Stiplet STIP terminology for an **APPLET**.

STL Standard Tessellation Language or Stereo-Lithography, file formats that describe the surface geometry of three dimensional objects as a series of unstructured triangulated surfaces in a three-dimensional Cartesian coordinate system without any on color, texture, or other design attributes. There are both **ASCII** and binary file representations, binary files are more common since they are more compact.

Stock Keeping Unit an identifier used by merchants to permit the systematic tracking of products and services at the level of each variation, bundle, service, fee or attachment.

Storage Area Network a technology for attaching computers and storage devices or storage systems that communicates over a network using block-based protocols. Similar to **NAS**.

Stormwater water that originates from precipitation events, snowmelt, or runoff from over watering.

STP SHIELDED TWISTED PAIR or SPANNING TREE PROTOCOL.

Streaming Media multimedia that is continuously displayed to the end-user while it is being delivered by the provider (see MEDIA SERVER). The term refers to the delivery method, not the media; streaming media systems frequently restrict off-line storage and access.

Structural Health Monitoring systems for monitoring building structure.

Structured Cabling see STRUCTURED CABLING SYSTEM.

Structured Cabling System (SCS), a standard hierarchical cable system infrastructure supporting multiple applications including voice, data, video, and BUILDING AUTOMATION SYSTEM; this has the advantage of reducing cabling costs and space requirements and facilitating integration. See standards TIA-568 for commercial, TIA-862 for building automation, TIA-942 for data centers, and Fig. 11 on page 287.

Structured Query Language a computer language for manipulating information stored in a relational database, also used as a synonym for relational database.

STS SHARED TENANT SERVICES or STATIC TRANSFER SWITCH.

Sub-Metering the ability to measure utility usage associated with a particular space or tenant within a facility.

Sub-metering/Billing measuring and charging for utilities, primarily electricity, at the tenant level.

Substantial Completion a milestone in the construction process that represents the point at which the client may request an occupancy permit in preparation for using the building and the start of any contractor warranties.

Sun-Blinds slatted, pleated or cellular devices used to control the amount of external light entering an area.

Sun Shade external architectural features that provide shade to the building. See also ROLLER-SHADE.

Super Audio Compact Disk a read-only optical disc designed to provide much higher fidelity digital audio than the original COMPACT DISC.

Supervised devices that are monitored in order to detect problems in communication or device function; supervised fire alarm devices are monitored by current sensing to verify the wired connection to the device. Contrast to ADDRESSABLE FIRE ALARM devices.

Supervisory Alarm a type of ALARM triggered by inability to communicate with a SUPERVISED device; this type of alarm is different from a smoke or fire alarm. For example, a break in the device wiring.

Supply Air the air that is delivered to the building by the HVAC system. Supply air is normally a mixture of RETURN AIR and OUTSIDE AIR.

Surface Acoustic Wave piezoelectric devices used in electronics circuits as filters, oscillators, transformers, or sensors based on the transduction of acoustic waves.

Surface Mount Technology a fabrication technique where components mount directly on printed circuit boards without inserting leads in holes or sockets.

Sustainable meeting the needs of the present without compromising the ability of future generations to meet their needs. Technologies, building materials, or business practices that use renewable resources, minimize environmental impacts, minimize energy usage, and are considered to be socially responsible.

SVB Switched Video Broadcast, see **Switched Digital Video**.

SVG (Scalable Vector Graphics), an open standard for describing two-dimensional vector graphics using XML. See www.w3.org/Graphics/SVG/ for more information.

Swim Lane Diagram a form of process flowchart that shows who is responsible for each activity or subset of the process. Parallel lines divide the chart horizontally or vertically into lanes, with one lane for each person or group.

Switch 1) an *electrical* switch provides the ability to connect or disconnect one or more circuits, there are also many specialized electrical switches, see for example **Static Transfer Switch**, **Limit Switch**; 2) a *telephone* switch or *circuit* switch provides the ability to make selective connections between on or more inputs and outputs; 3) a *network* switch, such as an Ethernet switch, connects together multiple physical networks (or segments) into a larger network by selectively forwarding packets or frames while providing isolation between network segments, a switch may also provide the ability to monitor and manage network links; 4) a *video* switch or *audio/video* switch enables selective connection of multiple sources to one or more displays or recorders, see also **Switched Digital Video** and **KVM** switch.

Switched Digital Video (SDV), techniques for efficiently distributing digital video via a network infrastructure with limited capacity. In **hybrid fiber coax** networks all channels are delivered to the fiber optic node which dynamically assigns channels on the coax portion of the network based on user requests, thus makes more efficient use of the limit number of coax channels by only transmitting channels that are being watched. **IPTV** and other **IP** based video services may use **unicast** or **multicast streaming media** protocols such as **RTSP**.

SXRD Silicon Crystal Reflective Display, an advanced form of **LCoS** that offers high resolution and no lines between pixels by using thin vertical liquid crystals to block light from a reflective aluminum substrate.

Syad (Systems Administrator) or **Syad**, a person or group responsible for maintaining and operating a computer system or network (see also **NETWORK ADMINISTRATOR**). Responsibilities may include installing, supporting, and maintaining computer system hardware and software, planning for and responding to outages and other problems, establishing and administering access control policies, measuring and monitoring system resource utilization, and planning future resource needs.

Symmetric Digital Subscriber Line a form of **DSL** offering the same transmission speeds in both directions.

Symmetrical Multiprocessing a type of computer system architecture with two or more identical processors (**CPUs**) and shared memory.

Synchronized Multimedia Integration Language is a markup language for multimedia presentations based on **XML** that defines timing, layout, animations, visual transitions, and media embedding, etc. For more information see www.w3.org/AudioVideo/.

Synchronous Data Link Control a communication protocol for layer two of the **OSI MODEL** developed by IBM; later became the basis for **HDLC**.

Synchronous Data Transmission data transmission where the sending and receiving devices are synchronized based on a clock signal provided by the transmission equipment. Synchronous transmission is more complicated and more efficient than **ASYNCHRONOUS DATA TRANSMISSION**.

Synergy cooperative action between two or more systems or groups that produces a result that is greater than the sum of the individual contributions.

System Integrator (SI), a person or organization that specializes in bringing together multiple components, frequently from different vendors, and configuring them to work together as a single system.

Systems Administrator or **Syad,** a person or group responsible for maintaining and operating a computer system or network (see also **Network Administrator**). Responsibilities may include installing, supporting, and maintaining computer system hardware and software, planning for and responding to outages and other problems, establishing and administering access control policies, measuring and monitoring system resource utilization, and planning future resource needs.

Système International d'unités or **International System of Units,** the modern form of the metric system of international standard measures.

T

T&E Travel and Entertainment, a type of credit card.

T/C (Thermocouple) a general class of electronic devices that can be used to measure temperature or to generate small amounts of electricity.

T1 T-1, or **DS1,** a data circuit with a speed of 1.544 Mbit/s line rate.

TAB (Testing, Adjusting, and Balancing) one step in the process of **Commissioning** a HVAC system.

Tailgating a situation where someone follows an authorized person through a secured door or other entrance after the authorized person opens the door legitimately.

Tamper switch a sensor on a manual cutoff valve in a fire sprinkler system that provides a supervisory alarm signal to the fire alarm system if the valve closes.

Target Of Evaluation (TOE), Common Criteria terminology for the product or system that being evaluated.

Task Lighting localized lighting provided in a user's work area.

Task Tuning setting light levels and other environment parameters to suit the particular task or other use of a workspace.

TB (Terabyte) one trillion, or ten to the twelfth power, bytes (characters) of storage. See Fig. 1 on page 279.

TBP (Transaction Based Pricing) usage based charging for outsourcing, software usage, after-hours access, or other services.

TCB Telecommunications Backbone, another term for core network.

TCO (Total Cost of Ownership) all the costs involved in acquiring, supporting, and disposing of an asset or technology.

TCP (Transmission Control Protocol) a transport layer protocol in the Internet protocol suite that provides reliable connections and in-order delivery using the best effort delivery mechanisms provided by the **IP** layer.

TCP/IP shorthand for **TCP** over **IP**, these are the basic protocols of the Internet; this term is frequently used to mean the entire suite of Internet protocols formalized by the **IETF** in addition to TCP and IP, see Fig. 7 on page 283.

TCT (Total Cycle Time) the elapsed time from the beginning to the end of process as perceived by the customer.

TDEA (Triple Data Encryption Algorithm) a block cipher created by applying the **Data Encryption Standard** (DES) cipher three times with different keys.

TDMA Time Division Multiple Access, IS-136, or D-AMPS, a 2G digital cellular telephony standard.

TE (Telecommunications Enclosure) a case or housing dedicated to the telecommunication function and support facilities.

Technical and Office Protocol the administrative portion of the MAP/TOP protocol suite for computer integrated manufacturing circa 1980.

Technology Road Map or **Roadmap,** a strategic or tactical plan that examines currently available technologies, anticipated future developments, and gaps where additional technology developments are needed in order to meet a goal or objective.

Telco a slang term for any telephone company; may imply a legacy carrier.

Telcordia or **Telcordia Technologies,** a telecommunications R&D and standards development organization formerly known as Bell Communications Research or Bellcore. For more information see www.telcordia.com.

Telecommunications Enclosure a case or housing dedicated to the telecommunication function and support facilities.

Telecommunications Grounding Busbar (TGB), signal ground connection for telecommunication equipment, terminology used in **TIA-569** and related standards; see also **BUS.**

Telecommunications Industry Association organization that, among other things, facilitates the process of developing standards and submits them to ANSI for adoption as American National Standards and international standards.

Telecommunications Outlet (TO) or **Telecommunications Connector,** the user connection facility provided in a Work Area as part of a **STRUCTURED CABLING SYSTEM.** See also MUTOA the multiple work area version.

Telecommunications Room a dedicated and specially equipped space for telecommunications equipment and cabling. See for example, **TIA-569.**

Telecommunications Space (TS), general term covering EQUIPMENT ROOM, TELECOMMUNICATIONS ROOM, etc.

Telepresence technologies which allow a person to feel as if they were present, to give the appearance of being present, or to have an effect, at a location other than their actual location.

Telnet TELetype NETwork, a protocol for remote terminal connections over the Internet standardized as RFC 854. Telnet connections are considered to be a security risk; **SSH** is the secure equivalent.

Template a stencil, pattern. or file that serves as an example and provides guidance for the user or provides control over content and format.

Tenant Improvements construction and modification of a leased space to fit the needs of the tenant.

Tenant Metering SUB-METERING at the level of the individual tenant.

Terabit data transmission speed measured in trillions of bits per second, one terabit per second is the same as 1,000 gigabits per second. See Fig. 2 on page 279.

Terabyte one trillion, or ten to the twelfth power, bytes (characters) of storage. See Fig. 1 on page 279.

Testing, Adjusting, and Balancing one step in the process of COMMISSIONING a HVAC system.

Text Message a short alphanumeric message (160 characters or less) sent to a mobile device via **SHORT MESSAGE SERVICE**. Gateways enable text messages to and from e-mail, **INSTANT MESSAGING** services, etc.

Text-To-Speech (TTS), a hardware and software system for converting normal language text into speech also known as Speech Synthesis. Multiple techniques are used including systems that store entire words or sentences for a specific domain, systems that concatenate pieces of recorded speech, and systems that construct utterances out of component speech sounds (phones or diphones).

TFT (Thin Film Transistor) one technology used for liquid crystal displays.

TGB (Telecommunications Grounding Busbar), signal ground connection for telecommunication equipment, terminology used in **TIA-569** and related standards; see also **BUS**.

The Open Group a computer industry consortium founded through the merger of X/Open and the Open Software Foundation that sets standards for Unix and related software. See www.opengroup.org for more information.

Thermocouple a general class of electronic devices that can be used to measure temperature or to generate small amounts of electricity.

Thick Client in a **CLIENT-SERVER** application architecture a thick client is a hardware and software system that provides local processing and data storage capability in the user's device; contrast to **THIN CLIENT**.

Thin Client in a **CLIENT-SERVER** application architecture a thin client provides only limited local processing in the user device and does not store application data; contrast to **THICK CLIENT**. Thin clients may be limited function devices like a **NETWORK COMPUTER** or standard software like a **WEB BROWSER**.

Thin Film Transistor one technology used for liquid crystal displays.

Threshold Limit Value occupational exposure guidelines for chemical substances and physical agents.

TI (Tenant Improvements) construction and modification of a leased space to fit the needs of the tenant.

TIA (Telecommunications Industry Association) organization that, among other things, facilitates the process of developing standards and submits them to ANSI for adoption as American National Standards and international standards.

TIA-232 ANSI/TIA-232-F-1997 Interface Between Data Terminal Equipment and Data Circuit-Terminating Equipment Employing Serial Binary Data Interchange serial interface supporting SYNCHRONOUS DATA TRANSMISSION or ASYNCHRONOUS DATA TRANSMISSION.

TIA-568 or **TIA/EIA-568-B**, a set of standards for generic commercial building telecommunications cabling (ANSI/TIA/EIA-568-B.1-2001, -B.2-2001, and -B.3-2001). These standards cover structured cabling systems that can be used within buildings and between buildings in a campus environment and includes high-performance twisted pair cabling and fiber optic cables in a hierarchical star topology. ISO/IEC 11801:2002 is the international equivalent.

TIA-569 or **ANSI/TIA/EIA-569-B**, Commercial Building Standard for Telecommunications Pathways and Spaces specifies entrance facilities, service entrance pathways, equipment rooms, intra-building backbone pathways, telecommunications rooms, horizontal pathways, conduits, grounding and bonding.

TIA-606 or **ANSI/TIA/EIA-606-A**, Administration Standard for Telecommunications Infrastructure covers identifiers, records, reports, drawings, and work orders.

TIA-862 or **ANSI/TIA/EIA-862**, Building Automation Systems Cabling Standard for Commercial Buildings specifies generic cabling topology, architecture, design, installation practices, test procedures, and coverage areas to support building automation systems. Other LOW VOLTAGE systems including audio/video, paging, service/equipment alarms, voice/data communications, wireless access points, are also supported by the cabling infrastructure requirements of this Standard. This standard is essentially an extension of ANSI/TIA/EIA-568-B (see **TIA-568**) that provides additional features for Building Automation.

TIA-942 or ANSI/TIA/EIA-942, Telecommunications Infrastructure Standard for Data Centers covers site space and layout, cabling infrastructure (see **Structured Cabling System**), tiered reliability (see **Tier** and Fig. 5 on page 281), and environmental considerations. See Fig. 16 on page 292

TIE Technical Information Exchange, a program of the US Department of Energy. See www.em.doe.gov/tie/ for more information.

Tier a classification system for data centers defined by The Uptime Institute based on specific availability and reliability features; Tier I is lowest and IV is highest. See Fig. 5 on page 281, www.upsite.com, and the **TIA-942** standard.

TIFF or **TIF,** originally Tagged Image File Format, a file format for storing bitmap images together with color space and other information published by Adobe as an open standard.

Tilt to angle a camera or other device up and down.

Time and Attendance a system for tracking personnel attendance and time spent at work for payroll and other functions.

Time-Division Multiplexing a way to transmit two or more signals apparently simultaneously over one communication channel by physically taking turns on the channel and creating sub-channels based on assigned time slots.

Time Weighted Average a method for calculating permissible exposure to toxics or hazards, see **PEL.**

Timestamp 1) a time code assigned to a transaction, alarm, or log entry; 2) Unix time measured as the number of seconds since 00:00:00 **UTC** on January 1, 1970.

TL Truckload.

TLD (**Top Level Domain**) the right most portion of a **Domain Name** specifying which global registry defines the domain name. There are generic global registries such as .com, .org, etc. that are three or more characters in length and country-specific registries such as .us, .fr, etc. that are two characters in length and assigned based on ISO territory codes.

TLS (**Transport Layer Security**) a cryptographic protocol for secure communications, successor to **SSL**. See RFC 4346 for more information.

TLV (**Threshold Limit Value**) occupational exposure guidelines for chemical substances and physical agents.

TM Transportation Management.

TMDS (**Transition Minimized Differential Signaling**) is a technology for encoding high-speed serial data used by the DVI and HDMI video interfaces.

TMGB Telecommunications Main Grounding Busbar, the main **Telecommunications Grounding Busbar**.

TMS **Transaction Management System** or Transportation Management System.

TO (**Telecommunications Outlet**) or **Telecommunications Connector**, the user connection facility provided in a Work Area as part of a **Structured Cabling System**. See also **MUTOA** the multiple work area version.

TOE (**Target Of Evaluation**), **Common Criteria** terminology for the product or system that being evaluated.

TOG (**The Open Group**) a computer industry consortium founded through the merger of X/Open and the Open Software Foundation that sets standards for Unix and related software. See www.opengroup.org for more information.

Token Ring a LOCAL AREA NETWORK technology originally developed and promoted by IBM and standardized as IEEE 802.5.

Ton a unit for measuring the rate of heat transfer; one ton equals 12,000 BTUH.

TOP (Technical and Office Protocol) the administrative portion of the MAP/TOP protocol suite for computer integrated manufacturing circa 1980.

Top Level Domain the right most portion of a DOMAIN NAME specifying which global registry defines the domain name. There are generic global registries such as .com, .org, etc. that are three or more characters in length and country-specific registries such as .us, .fr, etc. that are two characters in length and assigned based on ISO territory codes.

Toplighting a DAYLIGHTING technique where natural light enters through an opening in the top floor of a building; it provides the most uniform distribution of light across the workplane and better visual comfort than SIDELIGHTING. Toplighting may be provided by skylights, clerestory windows, light boxes, sawtooth roof monitors, etc.

Topology a branch of mathematics dealing with the nature of space and structures. Network topology deals with the connections and possible paths between network nodes; see for example STAR NETWORK, MESH NETWORK, BUS, etc.

Total Cost of Ownership all the costs involved in acquiring, supporting, and disposing of an asset or technology.

Total Cycle Time the elapsed time from the beginning to the end of process as perceived by the customer.

Total Value of Ownership a measure used to compare investment alternatives that incorporates costs (see TCO), benefits, and intangible factors.

Total Volatile Organic Compounds air pollution from all sources, see VOLATILE ORGANIC COMPOUNDS.

TOU Time of Use.

Touch Screen or **Touchscreen,** a display with a finger tracking overlay that enables simultaneous display and input. Multiple types of finger tracking technology are available; some do not actually require the finger to contact the device.

Touchtone or **Touch-Tone,** a trademarked term for **DUAL TONE MULTI-FREQUENCY** telephone signaling.

TQC Total Quality Control.

TQM Total Quality Management.

TR (Telecommunications Room) a dedicated and specially equipped space for telecommunications equipment and cabling. See for example, **TIA-569.**

Traffic Management systems and processes for monitoring and managing the flow of people, goods, vehicles, or data in a congested environment.

Transaction 1) in *Real Estate* the term transaction is most commonly used to mean a purchase/sale although it can include leasing and other activities; 2) in *Information Technology* a transaction is a unit of work that includes all of the database updates required for a specific business activity and the requirement that either all updates are successful or all updates are backed out (see **ATOMIC TRANSACTION**), for example the credit card charges, accounting entries, and inventory updates for a sale.

Transaction Based Pricing usage based charging for outsourcing, software usage, after-hours access, or other services.

Transaction Management System or **Transaction Management Service** (TMS), an application that supports the negotiation and processing of Real Estate **TRANSACTIONS.**

Transaction Processing any application that processes business transactions for billing or other purposes, typically with a financial value. May be ONLINE TRANSACTION PROCESSING or BATCH PROCESSING.

Transceiver a device that combines both a transmitter and a receiver.

Transcode to transfer or convert information from one coding system to another.

Transducer any sensing device that converts a signal from one form to another. For example, a pressure transducer produces an electrical voltage that can be used to measure pressure.

Transfer Air air that passes from one ventilation zone to another due to HVAC system imbalances, the STACK EFFECT, or other causes.

Transition Minimized Differential Signaling is a technology for encoding high-speed serial data used by the DVI and HDMI video interfaces.

Transmission Control Protocol a transport layer protocol in the Internet protocol suite that provides reliable connections and in-order delivery using the best effort delivery mechanisms provided by the IP layer.

Transport Layer Security a cryptographic protocol for secure communications, successor to SSL. See RFC 4346 for more information.

Trap a type of message in SNMP version one used to report an alert or other asynchronous event about a managed subsystem to the network management system. Terminology was changed to Notifications in later versions of SNMP.

Triage the process of evaluating and categorizing demands in order to maximize the effective use of limited resources. Triage may be used in medical emergencies or to prioritize problem reports.

Trigeneration see COMBINED HEATING POWER AND COOLING.

Trigger 1) a *condition* or combination of conditions that initiates an ALARM or a change in control system function; 2) a *protocol message* designed to start an action. For example, trigger a power-loss alarm if input voltage falls below 90 volts or send a trigger to the elevator control to activate certain floor buttons.

Triple Data Encryption Algorithm a block cipher created by applying the DATA ENCRYPTION STANDARD (DES) cipher three times with different keys.

Triple Play a package of three services: voice telephony, Cable TV or other video service, and Internet access.

TRM Technology Road Map.

Troffer a slang term for a recessed fluorescent light (combination of trough and coffer).

Trombe Wall a PASSIVE SOLAR heat collector in the form of a sun-facing wall that provides thermal mass (stone, concrete, adobe or water tanks) combined with an air space, insulated glazing and vents.

Trouble or **Defect**, an indication of an automatically detected problem with a system. For example, a contaminated smoke detector or an electrical problem in an alarm circuit.

Trouble Ticket a reported problem to be managed and resolved or an application for managing problem reports.

TS Telecommunications Space.

TTS (**Text-To-Speech**), a hardware and software system for converting normal language text into speech also known as Speech Synthesis. Multiple techniques are used including systems that store entire words or sentences for a specific domain, systems that concatenate pieces of recorded speech, and systems that construct utterances out of component speech sounds (phones or diphones).

Tubular Skylight see **Light Pipe**.

Tunneling Protocol a communications protocol which encapsulates or wraps one protocol inside another.

TUV Technischer Uberwachungs-Verein or Technical Surveillance Association, an organization that provides product testing and certification services.

TVO (Total Value of Ownership) a measure used to compare investment alternatives that incorporates costs (see **TCO**), benefits, and intangible factors.

TVOC (Total Volatile Organic Compounds) air pollution from all sources, see **Volatile Organic Compounds**.

TVSS Transient Voltage Surge Suppressor.

TWA (Time Weighted Average) a method for calculating permissible exposure to toxics or hazards, see **PEL**.

TXT a text file or plain text file format containing information with little or no formatting (contrast to **Rich Text Format** or **HTML**); also used as shorthand for **Text Message**.

U

U-factor the rate at which a window, door, or skylight conducts non-solar heat flow in units of BTU/hr-ft²-°F. Lower U-factors are more energy-efficient.

UART Universal Asynchronous Receiver Transmitter, a hardware device that implements a serial communications interface supporting **Asynchronous Data Transmission**.

UAT (User Acceptance Testing) a process where the customer verifies the expected function of a system or facility before accepting delivery or accepting transfer of ownership.

UBS (Uninterruptible Battery Supply) a direct current generator driven by an internal combustion engine that provides backup power to an UNINTERRUPTIBLE POWER SUPPLY or DC powered units.

UCD (Uniform Call Distributor) another term for AUTOMATIC CALL DISTRIBUTOR with the emphasis on equal workloads for all representatives.

UDDI (Universal Description, Discovery, and Integration) a standard XML-based API to a registry of network WEB SERVICES. See www.uddi.org for more information.

UDP (User Datagram Protocol) a protocol in the TCP/IP protocol suite that enables the efficient transmission of single packets or datagrams. See RFC 768 for more information.

UEM Unified Enterprise Management.

UFAD (Under Floor Air Distribution) a type of HVAC system where room air distribution uses a PLENUM located under a raised floor. Contrast to CFAD.

UGR (Unified Glare Rating) a standard definition for discomfort glare (the absence of VISUAL COMFORT) formalized as CIE-117 [1995].

UHF (Ultra High Frequency) radio band extending from 300 MHz to 3 GHz used for cellular telephones, paging, TV broadcast, wireless microphones, and other applications.

UI (User Interface) a general term covering all the ways in which people (the users) interact with a device or system including the types of information provided by the device and the ways in users exert control or provide input.

UL (Underwriters Laboratories) a product compliance and certification organization, see www.ul.com for more information.

UL Central Station an alarm system central station that is UL listed or certificated, see www.ul.com/alarmsystems/ for more information.

ULI Urban Land Institute, an education and research organization focused on the use of land, see www.uli.org for more information.

Ultra High Frequency radio band extending from 300 MHz to 3 GHz used for cellular telephones, paging, TV broadcast, wireless microphones, and other applications.

Ultra Mobile Broadband (UMB), a **3GPP2** project to develop next generation mobile phone standards based on **CDMA2000** with a goal of supporting peak rates of up to 280 Mbit/s.

Ultra Mobile Personal Computer (UMPC), a small form factor tablet PC specification developed by Microsoft, Intel, Samsung, and others. The goal is a compact device that is inexpensive and has good battery life.

Ultra-Wideband in general any radio technology with bandwidth larger than 500 MHz; in specific unlicensed use of 3.1–10.6 GHz for **WiMax** and other technologies.

Ultrasonic using sound with frequencies greater than the upper limit of human hearing (20,000 Hertz). Ultrasonic sensors are used as motion or occupancy sensors; ultrasound is also used for imaging, distance measurement, non-destructive testing, cleaning, etc.

Ultraviolet electromagnetic radiation with wavelengths shorter than that of visible light, but longer than X-rays.

UMB (Ultra Mobile Broadband), a **3GPP2** to project to develop next generation mobile phone standards based on **CDMA2000** with a goal of supporting peak rates of up to 280 Mbit/s.

UML (Unified Modeling Language) a specification technique used for software applications, business processes and data structures. The UML specification is maintained by the **OMG**, see www.uml.org for more information.

UMPC (Ultra Mobile Personal Computer), a small form factor tablet PC specification developed by Microsoft, Intel, Samsung, and others. The goal is a compact device that is inexpensive and has good battery life.

UMTS (Universal Mobile Telecommunications System) or **3GSM**, a third-generation mobile phone technology based on **GSM** and **W-CDMA**.

Under Floor Air Distribution a type of **HVAC** system where room air distribution uses a PLENUM located under a raised floor. Contrast to **CFAD**.

Underfloor Heating a form of RADIANT HEATING based on heating elements built into the floor, typically either electrical resistance wire or pipes that circulate warm water or other fluid.

Underwriters Laboratories a product compliance and certification organization, see www.ul.com for more information.

Undetected Problem Buildings buildings with defects or problems that have not yet been detected.

Unicast sending information to a single destination only. Contrast to MULTICAST.

Unicode a standard designed to represent text and symbols from all languages. See www.unicode.org for more information.

Unified Glare Rating a standard definition for discomfort glare (the absence of VISUAL COMFORT) formalized as **CIE-117** [1995].

Unified Messaging technology to consolidate voice mail, e-mail, fax and other messages into a single in-box that may be accessed through a variety of devices.

Unified Modeling Language a specification technique used for software applications, business processes and data structures. The UML specification is maintained by the **OMG**, see www.uml.org for more information.

Uniform Call Distributor another term for **Automatic Call Distributor** with the emphasis on equal workloads for all representatives.

Uniform Resource Identifier a way of identifying an information resource and optionally locating the resource on the Internet. Subtypes of URI include **URL** and **URN**.

Uniform Resource Locator one type of **URI** that provides a standard way to identify a resource on the Internet based on the access scheme (typically **HTTP**) and a scheme-specific address that typically includes the **Domain Name**, or an IP address, optionally followed by path information.

Uniform Resource Name a type of **URI** that provides a way of identifying an information resource without describing the location or address on the Web.

UniFormat a **Construction Specifications Institute** publication describing a Uniform Classification System for organizing preliminary construction information into a standard order or sequence on the basis of functional systems. For more information see www.csinet.org.

Uninterruptible Battery Supply a direct current generator driven by an internal combustion engine that provides backup power to an **uninterruptible power supply** or DC powered units.

Uninterruptible Power Supply a device that maintains a continuous supply of electric power by supplying power from a separate source, such as a battery, when utility power is not available.

United States Pharmacopeia a compendium of quality control tests for active and inactive medical ingredients published by the United States Pharmacopeial Convention. The initials USP are used to indicate materials that conform to the USP specifications and may be used medicinally.

Universal Description, Discovery, and Integration a standard XML-based **API** to a registry of network **WEB SERVICES**. See www.uddi.org for more information.

Universal Mobile Telecommunications System or 3GSM, a third-generation mobile phone technology based on **GSM** and **W-CDMA**.

Universal Plug and Play (UPnP), is a set of protocols promulgated by the **UPNP FORUM** to simplify the implementation of networks in the home and corporate environments. UPnP device control protocols are built upon open, Internet-based communication standards.

Universal Product Code standard barcode for retail merchandise administered by GS1. See www.gs1.org for more information.

Universal Serial Bus cabling and communications standard for connecting and optionally providing power to multiple devices. USB data rates may be 1.5, 12, or 480 Mbit/s.

Unix a computer **OPERATING SYSTEM** that originated at Bell Labs and is widely used for many types of applications. Several different groups have been responsible for Unix standards, the most recent is **THE OPEN GROUP**.

Unshielded Twisted Pair a type of wire used in communications cables that has two or more conductors that are systematically wrapped around each other and does not have an outer shield.

UPB (Undetected Problem Buildings) buildings with defects or problems that have not yet been detected.

UPC (Uniform Product Code) standard barcode for retail merchandise administered by GS1. See www.gs1.org for more information.

Uplink data transmission from the user device to the network, server, or satellite.

Upload to transfer data from the user's client computer or device to a server or network system.

UPnP (Universal Plug and Play), is a set of protocols promulgated by the UPnP Forum to simplify the implementation of networks in the home and corporate environments. UPnP device control protocols are built upon open, Internet-based communication standards.

UPnP Forum an industry initiative designed to enable simple and robust connectivity among consumer electronics, intelligent appliances and mobile devices from many vendors. For more information see www.upnp.org.

UPS (Uninterruptible Power Supply) a device that maintains a continuous supply of electric power by supplying power from a separate source, such as a battery, when utility power is not available.

URI (Uniform Resource Identifier) a way of identifying an information resource and optionally locating the resource on the Internet. Subtypes of URI include **URL** and **URN**.

URL (Uniform Resource Locator) one type of **URI** that provides a standard way to identify a resource on the Internet based on the access scheme (typically **HTTP**) and a scheme-specific address that typically includes the **Domain Name,** or an IP address, optionally followed by path information.

URN (Uniform Resource Name) a type of **URI** that provides a way of identifying an information resource without describing the location or address on the Web.

US Green Building Council (USGBC), a coalition of building industry leaders working to promote environmentally responsible buildings that are profitable and healthy places to live and work. For more information see www.usgbc.org; see also **LEED**.

Usability a quality attribute that assesses the ease-of-use of any technology. Usability also refers to methods for improving ease-of-use during design and development.

USART Universal Synchronous/Asynchronous Receiver Transmitter, a hardware device that implements a serial communications interface capable of supporting either SYNCHRONOUS DATA TRANSMISSION or ASYNCHRONOUS DATA TRANSMISSION.

USB (**Universal Serial Bus**) cabling and communications standard for connecting and optionally providing power to multiple devices. USB data rates may be 1.5, 12, or 480 Mbit/s.

Use Case a technique for capturing and documenting functional requirements. Each use case provides one or more scenarios that explain how the system should interact with users (also called actors) to achieve a business goal or function.

Usenet USEr NETwork, one of the oldest computer network discussion systems still in widespread use. Historical and current discussions may be accessed via Google Groups (groups.google.com).

User Acceptance Testing a process where the customer verifies the expected function of a system or facility before accepting delivery or accepting transfer of ownership.

User Datagram Protocol a protocol in the **TCP/IP** protocol suite that enables the efficient transmission of single packets or datagrams. See RFC 768 for more information.

User Interface a general term covering all the ways in which people (the users) interact with a device or system including the types of information provided by the device and the ways in users exert control or provide input.

USGBC (US Green Building Council), a coalition of building industry leaders working to promote environmentally responsible buildings that are profitable and healthy places to live and work. For more information see www.usgbc.org; see also **LEED**.

USP (United States Pharmacopeia) a compendium of quality control tests for active and inactive medical ingredients published by the United States Pharmacopeial Convention. The initials USP are used to indicate materials that conform to the USP specifications and may be used medicinally.

UT (Ultrasonic) using sound with frequencies greater than the upper limit of human hearing (20,000 Hertz). Ultrasonic sensors are used as motion or occupancy sensors; ultrasound is also used for imaging, distance measurement, non-destructive testing, cleaning, etc.

UTC Coordinated Universal Time, the zero point reference for time-zone offsets.

Utility Computing a flexible business model where computer resources are provided on-demand and charges are based on actual use.

UTP (Unshielded Twisted Pair) a type of wire used in communications cables that has two or more conductors that are systematically wrapped around each other and does not have an outer shield.

UV (Ultraviolet) electromagnetic radiation with wavelengths shorter than that of visible light, but longer than X-rays.

UWB (Ultra-Wideband) in general any radio technology with bandwidth larger than 500 MHz; in specific unlicensed use of 3.1–10.6 GHz for **WiMax** and other technologies.

V

Vacuum or **Partial Vacuum,** an area of low pressure compared to the gas pressure in the surrounding area (if it is pressurized) or below atmospheric pressure.

Vacuum Fluorescent a type of display technology used in electronic and automotive applications.

Value Added Network a communications network that provides additional services such as protocol conversion or message storage.

Value Added Reseller a company that adds features or services to the products they sell.

Value Added Services additional services that increase the value of the core service. For example, push-to-talk is a value added service for wireless telephones that makes it easier to reach frequently called numbers.

Value Engineering a process for identifying potential cost savings in design or construction by examining the ratio of function to cost.

Valve Regulated Lead Acid a type of rechargeable battery that does not require adding water. Contrast to VENTED LEAD ACID battery.

Vampire Load electrical power used by a device that is switched off.

VAN VALUE ADDED NETWORK or VIDEO ADVERTISING NETWORK.

VANet Vehicular Ad-Hoc Network, a WIRELESS AD-HOC NETWORK designed for communications among vehicles and between vehicles and roadside equipment.

VAR (Value Added Reseller) a company that adds features or services to the products they sell.

Variable Air Volume an air handler or other device that can provide different levels of air flow.

Variable Bit Rate a video two-pass encoding technique that analyzes and compresses images to an optimal data rate based on content, as opposed to a uniform data rate.

Variable Frequency Drive a device for controlling the speed of an electric motor by changing the frequency of its input power.

Variable Load Shedding the automatic reduction of peak period electrical demand from a building by turning equipment off or reducing operating levels.

Variable Refrigerant Flow an energy efficient technology for both heating and cooling based on distributed heat pump technology.

Variable Refrigerant Volume trade name for a **VARIABLE REFRIGERANT FLOW** technology.

Variable Voltage, Variable Frequency a type of variable speed motor controller that can adjust both voltage and frequency.

Varifocal a lens capable of operating at multiple focal lengths or distances; contrast to fixed-focus.

VAS (Value Added Services) additional services that increase the value of the core service. For example, push-to-talk is a value added service for wireless telephones that makes it easier to reach frequently called numbers.

VAV (Variable Air Volume) an air handler or other device that can provide different levels of air flow.

VBOC Virtual Building Operations Center, see **BUILDING OPERATIONS CENTER**.

VBR (Variable Bit Rate) a video two-pass encoding technique that analyzes and compresses images to an optimal data rate based on content, as opposed to a uniform data rate.

VBX (Visual Basic Extension) or Custom Control, a mechanism for packaging software components in Microsoft Visual Basic.

VCP Visual Comfort Probability.

VCSEL (Vertical-Cavity Surface-Emitting Laser) a type of semiconductor laser diode where the beam is emitted from the top surface. This technology promises cost and productivity benefits compared to conventional edge-emitting semiconductor lasers which emit from surfaces formed by cutting the chip out of a wafer.

VDT (Video Display Terminal), general term for any computer display devices including both **CRT** and **LCD** displays.

VDU (Visual Display Unit) a terminal, workstation, or other user interface device with a screen.

VE (Value Engineering) a process for identifying potential cost savings in design or construction by examining the ratio of function to cost.

Vector Graphics images described using computer algorithms to define the shapes, lines, animation, etc. Vector graphics can be smaller, sharper and more flexible than BITMAP graphics but may require more processing power.

Velocity Pressure the kinetic pressure in the direction of flow necessary to cause a gas at rest to flow with a given velocity.

Vendor Managed Inventory a business model where the supplier takes responsibility for maintaining product inventories in stores or other facilities.

Vented Lead Acid a type of battery, also known as flooded or wet cell batteries. Contrast to VALVE REGULATED LEAD ACID battery.

Ventilation Comfort Index an interior COMFORT INDEX.

Verification 1) to confirm or test the truth or accuracy of something; 2) in *testing* (see VERIFICATION AND VALIDATION) verification checks for conformance to specifications; 3) *software* verification involves formally proving that a program does exactly what is stated in the specification; 4) *data entry* verification typically involves entering data twice as a way to protect against errors.

Verification and Validation (V&V), checking that a system meets specifications and fulfils its intended purpose. Verification ensures that the system matches the original design; validation checks that the system fits the intended usage.

Vertical-Cavity Surface-Emitting Laser a type of semiconductor laser diode where the beam is emitted from the top surface. This technology promises cost and productivity benefits compared to conventional edge-emitting semiconductor lasers which emit from surfaces formed by cutting the chip out of a wafer.

Vertical Real Estate 1) real estate as a vertical market; 2) space above a building that may be used for antennas, other structures, SOLAR PANELS, etc.; 3) space on the exterior surfaces of a building that may be used for signage or other purposes.

Very Early Warning Smoke Detector a type of air-sampling smoke detection system that provides very early warning by detecting smoke particles in the incipient stage of a fire.

Very High Frequency radio band extending from 30 MHz to 300 MHz used for **FM** radio, TV, aircraft communications, wireless microphones and other applications.

Very Large Scale Integration integrated circuits containing large numbers of components.

VESA (Video Electronics Standards Association) an international group of video display manufacturers that helps establish standards. See www.vesa.org for more information.

VESDA trade name for one brand of VERY EARLY WARNING SMOKE DETECTOR.

VEWSD (Very Early Warning Smoke Detector) a type of air-sampling smoke detection system that provides very early warning by detecting smoke particles in the incipient stage of a fire.

VF (Vacuum Fluorescent) a type of display technology used in electronic and automotive applications.

VFD (Variable Frequency Drive) a device for controlling the speed of an electric motor by changing the frequency of its input power.

VGA (Video Graphics Array) an analog color display standard with 640×480 pixels.

VHF (Very High Frequency) radio band extending from 30 MHz to 300 MHz used for **FM** radio, TV, aircraft communications, wireless microphones and other applications.

VICS (Voluntary Interindustry Commerce Solutions Association) a retail industry supply chain standards organization. See www.vics.org for more information.

Video Advertising Network (VAN), a service that places and delivers video ads on Web sites and other media.

Video Display Terminal (VDT), general term for any computer display devices including both **CRT** and **LCD** displays.

Video Electronics Standards Association an international group of video display manufacturers that helps establish standards. See www.vesa.org for more information.

Video Graphics Array an analog color display standard with 640×480 pixels.

Video On Demand ability to replay stored video content on request.

Video Server a **MEDIA SERVER** specifically designed for video. Depending on the application, the video server may include special features for DIGITAL RIGHTS MANAGEMENT, scheduled broadcasts, combining or inserting video clips (leaders, trailers, commercials), watermarking, etc.

Video Telephony telephone service that delivers both audio and video.

Videoconferencing technology for meetings with two or more people at different locations where participants may see and hear each other.

VideoGuard a digital encryption system for **CONDITIONAL ACCESS** and **DIGITAL RIGHTS MANAGEMENT** developed by NDS. For more information see www.nds.com.

View Window see **VISION GLAZING**.

Virtual Concierge automated guest services either through a self-service device or video-telephony with a live person.

Virtual Local Area Network the ability to create logically independent and isolated **LAN** networks from a shared physical network by using switches and protocols that label or tag packets exchanged between switches with additional identifying information. For example, a shared building network could be partitioned into a VLAN for the **BUILDING AUTOMATION SYSTEM** and a separate VLAN for tenants giving each enhanced security and privacy. VLAN protocols include IEEE 802.1Q, Cisco ISL (Inter-Switch Link), and 3Com VLT (Virtual LAN Trunk).

Virtual Private Network a private communications network created using parts of a public or shared network.

Virtual Property Manager systems that enable the delivery of services like concierge, leasing, management, etc. from a centralized location using video networking.

Virtual Router Redundancy Protocol (VRRP), is a standards-based redundancy protocol designed to increase the availability of the gateway servicing hosts on the same subnet, described in IETF standard RFC 3768. See also Cisco **Hot Standby Router Protocol**.

Virtual Storage Area Network a technology for creating more than one **SAN** using shared or combined resources.

Virtualization technology that combines and partitions resources and creates the appearance of multiple virtual resources that operate independently. For example, server virtualization allows a single large server to run multiple **operating system** images simultaneously and do the work of several small servers.

Virulent dangerous even in small quantities; for example, a chemical that is highly toxic or a biological agent that is potentially fatal or debilitating.

Visible Notification Appliance a **notification appliance** that alerts by the sense of sight.

Visible Transmittance (VT), a number between 0 and 1 that represents the fraction of the visible spectrum of sunlight (380 to 720 nanometers), weighted by the sensitivity of the human eye, that is transmitted through a glazing material.

Vision Glass see **Vision Glazing**.

Vision Glazing windows, or the portion of larger windows, that provide a connection to the outdoors, typically vertical windows between 2.5 ft and 7.5 ft above the floor. Also known as **View Window**, contrast to **Daylight Glazing**.

Visual Basic Extension or Custom Control, a mechanism for packaging software components in Microsoft Visual Basic.

Visual Comfort subjective perception of the suitability of lighting taking into account uniform illumination, optimal light levels, glare, contrast, correct colors, and the absence of stroboscopic effect or intermittent light. See also **VCP**.

Visual Display Unit a terminal, workstation, or other user interface device with a screen.

Visualization techniques for creating images, diagrams, and moving pictures to communicate a message or to aid in the understanding of complex information.

Vivarium an area with specialized environmental controls and features to support keeping and raising animals or plants for observation or research.

VLA (Vented Lead Acid) a type of battery, also known as flooded or wet cell batteries. Contrast to **Valve Regulated Lead Acid** battery.

VLAN (Virtual Local Area Network) the ability to create logically independent and isolated **LAN** networks from a shared physical network by using switches and protocols that label or tag packets exchanged between switches with additional identifying information. For example, a shared building network could be partitioned into a VLAN for the **Building Automation System** and a separate VLAN for tenants giving each enhanced security and privacy. VLAN protocols include IEEE 802.1Q, Cisco ISL (Inter-Switch Link), and 3Com VLT (Virtual LAN Trunk).

VLSI (Very Large Scale Integration) integrated circuits containing large numbers of components.

VMI (Vendor Managed Inventory) a business model where the supplier takes responsibility for maintaining product inventories in stores or other facilities.

VOC (Volatile Organic Compounds) carbon based chemicals that can vaporize and enter the air. For example, paint thinners, solvents, petroleum based fuels, etc.

VOD (Video On Demand) ability to replay stored video content on request.

Voice/Data a system or network that supports both voice and data in digital form.

Voice Over IP a technology for providing voice telephony services using Internet Protocol and data networks as the transmission media.

VOIP (Voice Over IP) a technology for providing voice telephony services using Internet Protocol and data networks as the transmission media.

Volatile Organic Compounds carbon based chemicals that can vaporize and enter the air. For example, paint thinners, solvents, petroleum based fuels, etc.

Voluntary Interindustry Commerce Solutions Association a retail industry supply chain standards organization. See www.vics.org for more information.

VP (Velocity Pressure) the kinetic pressure in the direction of flow necessary to cause a gas at rest to flow with a given velocity.

VPM (Virtual Property Manager) systems that enable the delivery of services like concierge, leasing, management, etc. from a centralized location using video networking.

VPN (Virtual Private Network) a private communications network created using parts of a public or shared network.

VRF (Variable Refrigerant Flow) an energy efficient technology for both heating and cooling based on distributed heat pump technology.

VRLA (Valve Regulated Lead Acid) a type of rechargeable battery that does not require adding water. Contrast to VENTED LEAD ACID battery.

VRRP (Virtual Router Redundancy Protocol), is a standards-based redundancy protocol designed to increase the availability of the gateway servicing hosts on the same subnet, described in IETF standard RFC 3768. See also Cisco **Hot Standby Router Protocol**.

VRV (Variable Refrigerant Volume) trade name for a **Variable Refrigerant Flow** technology.

VSAN (Virtual Storage Area Network) a technology for creating more than one **SAN** using shared or combined resources.

VSD Variable Speed Drive.

VT Visible Transmittance

VVI Variable Voltage Inverter.

VVVF (Variable Voltage, Variable Frequency) a type of variable speed motor controller that can adjust both voltage and frequency.

W

W-CDMA Wideband Code Division Multiple Access, a 3G cellular telephony standard based on **CDMA**.

WA Work Area.

WAN (Wide Area Network) any network that extends over long distances. Contrast to **Local Area Network** or **Metropolitan Area Network**.

WanderGuard an **active RFID** tracking application designed to prevent persons at risk from leaving a facility unless they are accompanied. The system tracks the person using a wrist or ankle band and automatically locks doors or alarms if the person moves outside a defined area without being accompanied by an authorized person.

WAO Work Area Outlet.

WAP WIRELESS ACCESS POINT or WIRELESS APPLICATION PROTOCOL.

War Dialing a brute force technique for locating modems or fax machines by calling all the numbers in a given range.

War Driving wireless network equivalent of WAR DIALING, cruising around looking for wireless network signals and unprotected networks.

WAV Waveform audio format, audio file format standard typically used to store uncompressed audio in the pulse-code modulation format.

Wavelength Division Multiplexing (WDM), a technology for increasing the transmission capacity of a fiber optic link by using lasers that operate on different frequencies or colors of light.

WBEM (Web Based Enterprise Management) a DMTF systems management architecture designed to unify the management of enterprise computing environments. These standards include the COMMON INFORMATION MODEL COMPUTING.

WBP Whole Building Power.

WBS (Work Breakdown Structure) a hierarchical plan of tasks and subtasks required to complete a project showing the dependencies between tasks and resource needs.

WC Water Column, used to measure air pressure.

WDM (Wavelength Division Multiplexing), a technology for increasing the transmission capacity of a fiber optic link by using lasers that operate on different frequencies or colors of light.

WE Water Efficiency.

Web short for WORLD WIDE WEB, one of the applications that uses the Internet.

Web 2.0 so called second-generation **Web** based communities and social networking sites that facilitate collaboration and sharing between users (see for example **Wiki** technology). This reflects a change in the ways developers have used the Web platform and where the hype is focused, not an update to the technical specifications.

Web Address the information needed to locate an information resource on the Internet, technically a **URL**.

Web Based Enterprise Management a **DMTF** systems management architecture designed to unify the management of enterprise computing environments. These standards include the **Common Information Model computing**.

Web Browser a software application that enables a user to display and interact with hyperlinked text, images, and other information on Web pages and communicates using **HTTP** protocol. Browsers display pages by interpreting **HTML** markup codes based on local user preference settings. Certain types of programs called plugins may run within the browser environment. Examples of personal computer browsers include Microsoft Internet Explorer, Mozilla Firefox, Apple Safari, Netscape, Opera, etc.

Web Cam a video camera that connects to the Internet or Web, either directly or through a computer.

Web Designer a person who creates the visual appearance, navigation, and interactive features of a **Web Site**.

Web Developer a person who does the technical work to create and maintain a **Web Site** including creating the **HTML** formatting for content; may also include **Cascading Style Sheets**, **content management**, programming, and **database**.

Web Page a document that is available on the **Web** and contains information that is encoded in **HTML**.

Web Radio audio broadcast using the Web.

Web Server 1) a computer *program* that accepts **HTTP** requests from **Web browser** clients and returns or serves HTTP responses and contents such as **Web page**s and linked images, sounds, etc; 2) computer *systems* used to run Web Server software.

Web Services standards and protocols for application-to-application interfaces over a **TCP/IP** network. The Web Service protocol stack includes Service Transport (**HTTP, SMTP, FTP,** etc.), **XML** Messaging (**XML-RPC, SOAP,** and **REST**), Service Description (**WSDL**), and Service Discovery (**UDDI**).

Web Services Description Language standard for describing the public interface to a specific web service. See www.w3.org/TR/wsdl for more information.

Web Services Distributed Management a standard **Web Services** architecture for managing distributed resources. See www.oasis-open.org/committees/wsdm/ for more information.

Web Services Interoperability Organization an industry organization that provides profiles, sample applications and testing tools to promote Web Services interoperability. See www.ws-i.org for more information.

Web Services Resource Framework an open framework for modeling and accessing stored data using **Web Services**. See www.globus.org/wsrf/ for more information.

Web Site or **Website,** a collection of documents available on the **Web**.

Webcast short for web broadcast, to send audio and/or video live over the Internet.

Weigand or **Weigand Wire,** a special type of magnetic media that is embedded in the cards used for access control applications.

WEP (Wired Equivalent Privacy) a scheme to secure **Wi-Fi** wireless networks. WEP was superseded by **WPA** in 2003 and by the full IEEE **802.11i** in 2004.

WFM (Workforce Management) applications for tracking and managing personnel, may include a variety of applications including TIME AND ATTENDANCE, payroll, human resources management, scheduling, etc.

WfSC Microsoft Windows for Smart Cards.

White Box Testing a testing approach that provides the testers with complete knowledge of the infrastructure to be tested. Contrast to BLACK BOX TESTING.

White List an ACCESS CONTROL LIST naming users or systems that are allowed access; contrast to BLACKLIST.

White Noise a random signal with equal power in all frequency bands that is used for SOUND MASKING and other applications.

Wi-Fi a brand name reverse engineered from Wireless Fidelity to make IEEE 802.11 based wireless local area networks more consumer friendly. See www.wi-fi.org for more information.

Wi-Fi Protected Access a security system for Wi-Fi wireless networks based on parts of 802.11i. WPA was an interim solution created in response to weaknesses found in WEP and was replaced by full 802.11i, also known as WPA2, in 2004.

Wibree a digital radio technology designed for very low power consumption and short range (10 m) transmission at speeds of 1 Mbit/s or less using the 2.4 GHz ISM BAND. For more information see www.wibree.com.

Wide Area Network any network that extends over long distances. Contrast to LOCAL AREA NETWORK or METROPOLITAN AREA NETWORK.

Wideband Telephony telephone service capable of transmitting 150 Hz to 6.3 KHz, as contrasted to 300 Hz to 3.4 KHz for traditional telephones or 20 Hz to 20 KHz for CD quality audio.

Wiki collaborative software that allows viewers to edit, expand, and change the content of a Web site.

Wikipedia an online encyclopedia using **WIKI** technology and created from user contributed content.

WiMAX Worldwide Interoperability for Microwave Access, technology for wireless broadband access per IEEE **802.16**. Uses MESH NETWORK technology over 2-11 or 10-60 GHz radio to provide data rates up to 70Mbit/s over several miles.

Windows or **Microsoft Windows**, a family of computer OPERATING SYSTEMs with versions for personal computers, servers, personal digital assistants, etc.

WIP Work In Process or Work in Progress.

Wire Management 1) *physical* equipment for supporting and organizing wires and cables; 2) *systems* and business processes for record keeping related to wiring and cables, also known as WIRING SYSTEMS ADMINISTRATION.

Wired Equivalent Privacy a scheme to secure WI-FI wireless networks. WEP was superseded by **WPA** in 2003 and by the full IEEE **802.11i** in 2004.

Wireless without wires, usually implies radio frequency communications but the term also includes optical and infrared communications.

Wireless Access Point a device that connects wireless devices together to form a network or connect to the wired network.

Wireless Ad-hoc Network a self-organizing network that uses wireless links. Typically this is a **MESH NETWORK** where all nodes can take multiple roles including forwarding messages to other nodes.

Wireless Application Protocol a protocol suite intended to give mobile devices access to Internet services. See www.openmobilealliance.org for more information.

Wireless Local Area Network a local area network implemented using wireless technology such as **Wi-Fi**.

Wireless Microphone a microphone that uses radio (typically in the **UHF** band) or infrared light for transmission.

Wireless Personal Area Network a wireless network connecting devices close to one person. WPAN technologies include **Bluetooth** and infrared (**IrDA**).

Wireless Sensor Network is a network that includes small sensor nodes (or motes) and one or more base stations (also called sinks), that collect the data from the sensor nodes. For example, see **ZigBee** and **802.15.4**.

Wireless Service Provider any organization that provides wireless communication services including mobile phone operators, or cellular telephone companies, and organizations that operate **Wi-Fi** or **WiMax** networks.

WirelessHART a wireless version of the **Highway Addressable Remote Transducer** protocol for process instrumentation. For more information see www.hartcomm.org.

WirelessMAN official name for IEEE **802.16**, known commercially as **WiMAX**.

Wiring Closet a small room used for telecommunications equipment and cabling.

Wiring Systems Administration business processes and computer systems for record keeping and management of telecommunications infrastructure wiring. See **TIA-606** standard.

Wizard a program designed to help the user set up and configure hardware or software by guiding them through a step-by-step process.

WLAN WIRELESS LOCAL AREA NETWORK

WM Warehouse Management.

WML WIKI Markup Language or Wireless Markup Language used with **WIRELESS APPLICATION PROTOCOL**.

WMS Warehouse Management System.

Work Breakdown Structure a hierarchical plan of tasks and sub-tasks required to complete a project showing the dependencies between tasks and resource needs.

Workflow 1) the *operational* aspect of a work process including task structured, who performs each task, their relative order, how activities are synchronized, how information flows, and how tasks are tracked; 2) computer *applications* designed to support and automate workflows.

Workforce Management applications for tracking and managing personnel, may include a variety of applications including **TIME AND ATTENDANCE**, payroll, human resources management, scheduling, etc.

Workplane Illuminance the amount of light from all sources falling on a work surface.

World Wide Web a system of interlinked **HYPERTEXT** documents and applications that are available on the **INTERNET** and may be accessed using a **WEB BROWSER** or other device.

WPA (Wi-Fi Protected Access) a security system for Wi-Fi wireless networks based on parts of 802.11i. WPA was an interim solution created in response to weaknesses found in WEP and was replaced by full **802.11i**, also known as WPA2, in 2004.

WPAN (Wireless Personal Area Network) a wireless network connecting devices close to one person. WPAN technologies include **BLUETOOTH** and infrared (**IrDA**).

WS-I (Web Services Interoperability Organization) an industry organization that provides profiles, sample applications and testing tools
to promote Web Services interoperability. See www.ws-i.org for more
information.

WSCN Wireless Sensor and Control Network, see **WIRELESS SENSOR
NETWORK.**

WSDL (Web Services Description Language) standard for describing the
public interface to a specific web service. See www.w3.org/TR/wsdl for
more information.

WSDM (Web Services Distributed Management) a standard **WEB
SERVICES** architecture for managing distributed resources. See
www.oasis-open.org/committees/wsdm/ for more information.

WSN (Wireless Sensor Network) is a network that includes small sensor
nodes (or motes) and one or more base stations (also called sinks), that
collect the data from the sensor nodes. For example, see **ZIGBEE** and
802.15.4.

WSP (Wireless Service Provider) any organization that provides wireless
communication services including mobile phone operators, or cellular
telephone companies, and organizations that operate **WI-FI** or **WIMAX**
networks.

WSRF (Web Services Resource Framework) an open framework
for modeling and accessing stored data using **WEB SERVICES.** See
www.globus.org/wsrf/ for more information.

WWAN Wireless **WIDE AREA NETWORK,** for example, **WIMAX.**

WWW (World Wide Web) a system of interlinked **HYPERTEXT** documents and applications that are available on the **INTERNET** and may be
accessed using a **WEB BROWSER** or other device.

WYSIWYG What You See Is What You Get, a style of user interface for desktop publishing where the on screen appearance mimics the final result.

X

X.500 a series of ITU networking standards for electronic directory services, see www.itu.int/rec/T-REC-X.500/en for details. See also **LDAP**.

X.509 an ITU standard for **PUBLIC KEY INFRASTRUCTURE** that specifies public key **DIGITAL CERTIFICATE** formats and a method for authenticating keys using a certification path.

X11 the communications protocol portion of **X WINDOWS**.

X Terminal a limited function device that provides an **X WINDOWS** user interface to remote applications.

X Windows a networking and graphical user interface display protocol standard originally developed for **UNIX** and adopted by **LINUX** and other systems. Note: in X terminology the display server supports the user interface and client is the remote application; more common usage is client for the user interface and server for the application.

XACML eXtensible Access Control Markup Language, an access control policy language implemented in XML and a processing model for applying the policies. See www.oasis-open.org for more information.

XBRL Extensible Business Reporting Language, a series of **XML**-based standards for reporting and exchanging business and financial information. Standards are governed by XBRL International, a nonprofit consortium, for more information see www.XBRL.org.

xDSL shorthand for all types of **DSL** including **ADSL** and **SDSL**.

Xeriscaping or **Xeroscaping**, landscaping designed to survive without supplemental irrigation and requiring little care.

XGA eXtended Graphics Array also known as Super **VGA**, an analog color display standard supporting 800x600 or 1024x768 pixels.

XHTML eXtensible Hyper Text Markup Language, the successor to **HTML** that is based on **XML**, XHTML has stricter rules that make documents easier to process automatically and display on different devices.

XIF eXternal Interface File, a document describing the network variables and configuration properties supported by a **LonTalk** device.

XML eXtensible Markup Language, a very flexible format for exchanging data between systems that is used by a wide variety of applications. See www.w3.org/XML/ for more information.

XML-RPC Remote Procedure Call protocol that uses **XML** to encode its data and **HTTP** as a transport protocol. Predecessor to **SOAP**, largely obsolete.

XMPP (eXtensible Messaging and Presence Protocol) an open XML-based protocol for near-real-time instant messaging and **PRESENCE INFORMATION**. Standard was based on Jabber technology and formalized as RFC 3920 and RFC 3921.

XSLT eXtensible Stylesheet Language Transformations, an **XML**-based language that defines a series of template rules that can be used to transform an XML document into another XML document or a different format.

XSS Cross-Site Scripting, a type of computer security vulnerability or attack that involves a script or other software code injected into a web page.

xvYCC shorthand for Extended YCC Colorimetry or IEC 61966-2-4, a standard that supports 1.8 times as many colors as HDTV and enables more accurate and vivid colors.

YB (Yottabyte) one septillion, or ten to the twenty-fourth power, bytes (characters) of storage. See Fig. 1 on page 279.

YCbCr family of color spaces used in video systems where Y is the luma component and Cb and Cr are the blue and red chroma components.

Yottabyte one septillion, or ten to the twenty-fourth power, bytes (characters) of storage. See Fig. 1 on page 279.

Z

Z Coordinate vertical location information provided as part of indoor wireless **LOCATION-BASED SERVICES** to notify emergency responders of the exact location.

ZAT Zone Air Temperature.

ZB **ZETTABYTE** or **ZONE BOX**.

ZC (ZigBee Coordinator) a device that acts as the root of a **ZIGBEE** network tree and may act as a bridge to other networks.

ZDA (Zone Distribution Area), an optional interconnection point in the horizontal cabling between the **HORIZONTAL DISTRIBUTION AREA** and the **EQUIPMENT DISTRIBUTION AREA**.

ZEB (Zero Energy Building) a structure designed to have a net energy consumption of zero over a typical year.

ZED (ZigBee End Device) a ZIGBEE network device that contains just enough functionality to talk to its parent node (a ZIGBEE COORDINATOR or a ZIGBEE ROUTER).

Zero-Day or **Zero-Hour**, threats that attack undisclosed computer system vulnerabilities. Zero-day attacks can be extremely dangerous because they exploit security holes for which no solution is currently available.

Zero Energy Building a structure designed to have a net energy consumption of zero over a typical year.

Zero VOC paints or other commodities formulated to eliminate VOLATILE ORGANIC COMPOUNDS.

Zettabyte one sextillion, or ten to the twenty-first power, bytes (characters) of storage. See Fig. 1 on page 279.

Zigbee suite of high level communication protocols for self-organizing MESH NETWORKS using small, low-power digital radios based on the IEEE 802.15.4 standard intended for embedded applications requiring low data rates and low power consumption for industrial control, sensing, smoke and intruder warning, building automation, home automation, etc.

ZigBee Coordinator a device that acts as the root of a ZIGBEE network tree and may act as a bridge to other networks.

ZigBee End Device a ZIGBEE network device that contains just enough functionality to talk to its parent node (a ZIGBEE COORDINATOR or a ZIGBEE ROUTER).

ZigBee Router a device that routes information within a ZIGBEE network.

Zip 1) a format for loss-less file compression that also allows multiple files or directories to be stored a single file; 2) Zip Drive or Zip Disk, removable media products from Iomega.

Zone Box a form of TELECOMMUNICATIONS ENCLOSURE serving one zone; various types are available including wall, ceiling, and in-floor units.

Zone Cabling cabling using a CONSOLIDATION POINT in the horizontal cable located in a ZONE BOX near the work area and shorter cables from the zone box to each work area. Contrast to HOME RUN CABLING.

Zone Distribution Area (ZDA), an optional interconnection point in the horizontal cabling between the HORIZONTAL DISTRIBUTION AREA and the EQUIPMENT DISTRIBUTION AREA.

Zoning 1) governmental land-use regulation; 2) the practice of providing independent heating or cooling controls for different areas within a structure; 3) spatial subdivisions for networking or control systems.

Zoom to adjust a camera or other device to show more detail in a specific area by changing the effective focal length of the lens or enlarging part of the image.

ZR (**ZigBee Router**) a device that routes information within a ZIGBEE network.

Figures

Fig. 1
Storage Capacity

SI Unit	Abbr	Common	Bytes	10^n	2^n
Kilobyte	kB	Thousand	1,000	10^3	2^{10}
Megabyte	MB	Million	1,000,000	10^6	2^{20}
Gigabyte	GB	Billion	1,000,000,000	10^9	2^{30}
Terabyte	TB	Trillion	1,000,000,000,000	10^{12}	2^{40}
Petabyte	PB	Quadrillion	1,000,000,000,000,000	10^{15}	2^{50}
Exabyte	EB	Quintillion	1,000,000,000,000,000,000	10^{18}	2^{60}
Zettabyte	ZB	Sextillion	1,000,000,000,000,000,000,000	10^{21}	2^{70}
Yottabyte	YB	Septillion	1,000,000,000,000,000,000,000,000	10^{24}	2^{80}

Fig. 2
Transmission Speeds

SI Unit	Abbr	Common	Bits per second	10^n
Kilobit	kbit/sec	Thousand	1,000	10^3
Megabit	Mbit/sec	Million	1,000,000	10^6
Gigabit	Gbit/sec	Billion	1,000,000,000	10^9
Terabit	Tbit/sec	Trillion	1,000,000,000,000	10^{12}

Fig. 3

Chemical Refrigerants

Examples of chemicals used as HVAC refrigerants and for other purposes
with their R-number, Ozone Depletion Potential (ODP), and Global
Warming Potential (GWP) values. ODP and GWP values are from "The
Treatment by LEED® of the Environmental Impact of HVAC Refrigerants"
by the TSAC HCFC Task Group, September 28, 2004, available from www.
usgbc.org/Docs/LEED_tsac/TSAC_Refrig_Report_Final-Approved.pdf
accessed on May 25, 2007.

R number	Chemical Name	Class-ification	Trade Name	ODP	GWP
R11	Trichlorofluoromethane	CFC-11	Freon 11	1.0	4,680
R12	Dichlorodifluoromethane	CFC-12	Freon 12, Halon 122	1.0	10,720
R22	Chlorodifluoromethane	HCFC-22	Freon 22	0.04	1,780
R23	Trifluoromethane	HFC-23	Freon 23	4×10^{-4}	12,240
R123	2,2-Dichloro-1,1,1-tri-fluoroethane	HCFC-123	SUVA 123	0.02	76
R134a	1,1,1,2-Tetrafluoroethane	HFC-134a		10^{-5}	1,300
R290	Propane		Duracool	0	3
R407C	Mix of R-32/R-125/R-134a (23/25/52)			10^{-5}	1,700
R410A	Difluoromethane and Pentafluoroethane	HFC-410A	Puron	10^{-5}	1,890
R717	Ammonia			0	0
R718	Water			0	0
R744	Carbon Dioxide			0	1

Fig. 4

Availability and Downtime

Availability Level	Percentage	Downtime per Year
Three Nines	99.9%	8 hours 46 minutes
Four Nines	99.99%	53 minutes
Five Nines	99.999%	5 minutes
Six Nines	99.9999%	32 seconds

Fig. 5

Data Center Tiers

	Tier I	Tier II	Tier III	Tier IV
Typical Availability	99.67%	99.75%	99.98%	99.99%
Annual IT Downtime	28.8 hours	22.0 hours	1.6 hours	0.8 hours
Concurrent Maintenance	No	No	Yes	Yes
Fault Tolerance	No	No	No	Yes
UPS Configuration	Capacity N	N+1 Redundant	Distributed Redundant	2N or 2N+1

Fig. 6

OSI Model

OSI Layer	Name	Function	Examples
7	Application	Network process to application	HTTP, SMTP, X.500
6	Presentation	Data representation and encryption	MIME, ASN.1, SSL
5	Session	Host-host communication	SIP
4	Transport	End-to-end connections and reliability	TCP, UDP
3	Network	Path determination and logical addressing	IP, IPsec
2	Data Link	Physical addressing	Ethernet, ATM, FDDI, PPP
1	Physical	Media, signal and binary transmission	RS-232, DSL, SONET

Fig. 7

Internet Protocol Stack

OSI Layer	TCP/IP Layer	Name	Example Protocols
7	5	Application	DHCP, DNS, FTP, HTTP, IMAP, IRC, NNTP, POP3, SMTP, SNMP, SOAP, TELNET, XMPP, etc.
6		Presentation	MIME, ASN.1, SSL
5		Session	SIP
4	4	Transport	TCP, UDP
3	3	Network	IP, IPv6, IPsec
2	2	Data Link	Ethernet, FDDI, PPP
1	1	Physical	RS-232, DSL, SONET

Fig. *8*

ITU Radio Bands

ITU band	Name	Abbr	Frequency Range	Uses
1	Extremely Low Frequency	ELF	3–30 Hz	Submarine communications
2	Super Low Frequency	SLF	30–300 Hz	Submarine communications
3	Ultra Low Frequency	ULF	300–3000 Hz	Communication within mines
4	Very Low Frequency	VLF	3–30 kHz	Submarines, avalanche beacons, heart rate monitors, geophysics
5	Low Frequency	LF	30–300 kHz	Navigation, time signals
6	Medium Frequency	MF	300–3000 kHz	AM broadcasts
7	High Frequency	HF	3–30 MHz	Shortwave and aviation communications
8	Very High Frequency	VHF	30–300 MHz	FM, television, and aircraft communications
9	Ultra High Frequency	UHF	300–3000 MHz	Television, microwave ovens, mobile phones, wireless LAN, Bluetooth, GPS and Two-Way Radios
10	Super High Frequency	SHF	3–30 GHz	Microwave devices, wireless LAN, radar
11	Extremely High Frequency	EHF	30–300 GHz	Radio astronomy

Fig. 9

Common Frequency Bands

Band	Usage	Frequency Ranges
Cell	Cellular telephony AMPS, CDMA, D-AMPS, GSM, TDMA	824-849, 869-894, 896-901, 935-940 MHz
PCS	Personal Communications Service (cellular telephony) CDMA, D-AMPS, GSM	1850-1910, 1930-1990 MHz
AWS	Advanced Wireless Services 3G	700, 1432-1435, 1710-1755, 2110-2170, 2500-2690 MHz
SMR	Specialized Mobile Radio SMR/Nextel, iDEN	806-824, 851-869 MHz
ISM	Industrial, Scientific, Medical	Includes 900 MHz, 2.4, 5.8 GHz
UWB	Ultra-Wide Band UWB	3.1–10.6 GHz

See Fig. 10 on page 286 for standards using these bands.

Fig. 10

Wireless Standard Summary

Standard	Cell/ PCS	AWS	ISM	Other	Mobile Gen	Max Rate	Range
1x, CDMA2000	Y	Y			2	0.307	~18 mi
EV-DO, CDMA2000	Y	Y			2	3.1	~18 mi
EDGE/ GPRS	Y				2	0.474	~18 mi
UMTS over W-CDMA	Y	Y			3	0.3	~18 mi
UMTS HSPDA	Y	Y			3	14	~18 mi
UMTS-TDD	Y	Y		Y	3	16	~18 mi
WiMAX 802.16e				Y		70	~4 mi
WiFi 802.11a			Y	Y		54	
WiFi 802.11b			Y			11	~30 m
WiFi 802.11g			Y			54	~30 m
WiFi 802.11n			Y			200	~50 m
Bluetooth 802.15.1			Y			2.1	1 to 100m
ZigBee 802.15.2			Y			.25	100m
EnOcean			Y			.12	300m
Wibree			Y			1.0	10m
Wireless USB, UWB				UWB			
NFC				HF		.1 - .4	< 0.2m
RFID				UHF		.001- 0.2	0.01- 10m

Fig. 11
Structured Cabling System

Structured Cabling provides a standard infrastructure to support multiple applications. The standards cover cabling and building spaces network structure but not the actual networking equipment.

Structured Cabling includes these subsystems as shown in Fig. 13 on page 289:

- Entrance Facilities that connect the building to carriers and other campus buildings.

- Equipment Rooms hold equipment for the building including the Main Cross-Connect for the core network.

- Telecommunications Rooms typically on each floor are the location of the Horizontal Cross-Connect linking the backbone or core network with the Horizontal Cabling and Edge Network.

- Backbone Cabling for the core network typically runs vertically within a building.

- Horizontal Cabling for the edge network extends out to the work areas.

- Work-Area Components connect end-user equipment.

TThere are multiple standards based on this framework including:

- TIA-568 Generic Cabling System shown in Fig. 14 on page 290 is the basic structured cabling system.

- TIA-862 Building Automation System Cabling shown in Fig. 15 on page 291 provides additional facilities to support building systems.

- TIA-942 Data Center Cabling shown in Fig. 16 on page 292 provides additional capabilities for data centers.

Cables are classified by category or Cat as shown in Fig. 12 on page 288. Some categories of cables are no longer allowed under TIA-568-B.

Fig. 12

Cable Categories

Cat	Usage	TIA/EIA-568-B
1	Telephone and doorbell wiring	N
2	4 Mbit/s Token Ring	N
3	10 Mbit/s Ethernet	Y
4	16 Mbit/s Token Ring	N
5	100 Mbit/s Ethernet, may be unsuitable for 1000BASE-T gigabit Ethernet.	N
5e	100 Mbit/s and Gigabit Ethernet	Y
6	Up to 250 MHz, more than double cat 5 and 5e.	Y
6a	Future specification for 10 Gbit/s	N
7	ISO/IEC 11801 Class F cabling	N

Fig. 13
TIA-569 Space Definitions

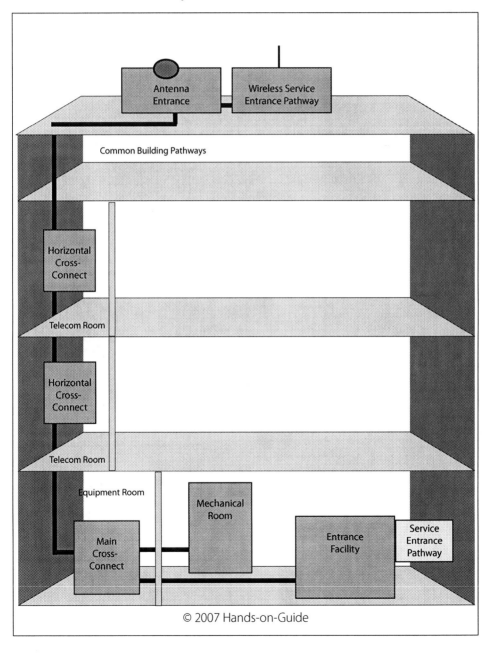

Fig. 14

TIA-568 Structured Cabling System

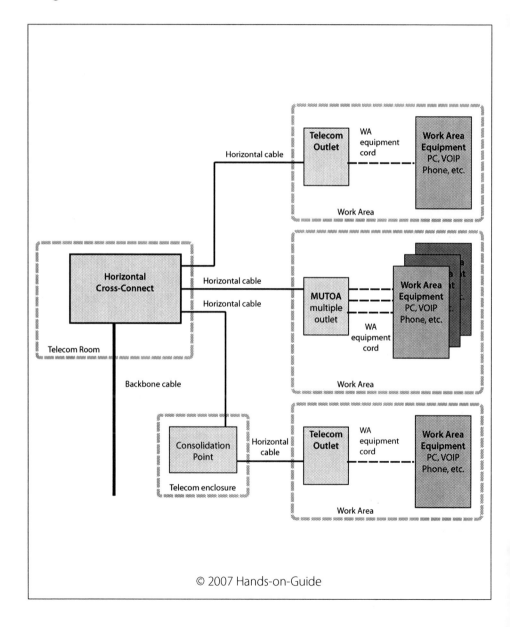

© 2007 Hands-on-Guide

Fig. 15
TIA-862 Building Automation System Cabling

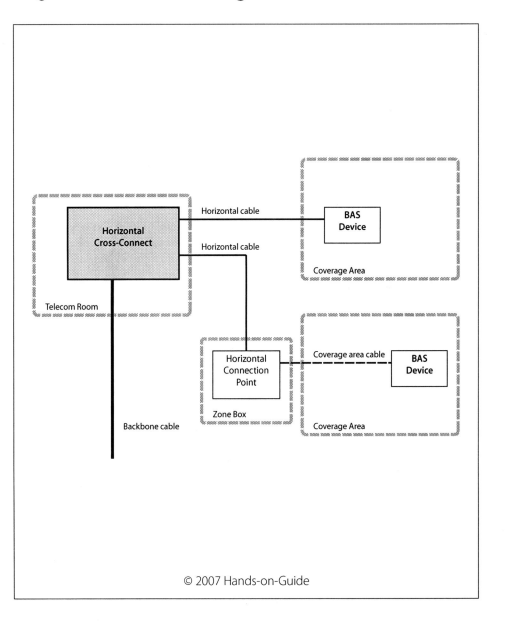

About the Authors

Building Intelligence Group consultants help building owners and managers, system integrators and suppliers profit from intelligent building technologies. Our services range from requirements definition and building systems design for specific projects to multi-project plans and automation strategies, market research and marketing strategy. We are frequent speakers at industry events and actively involved with industry organizations, Intelligent Building education, and standards development. For more information see www.buildingintelligencegroup.com.

Chuck Ehrlich is CTO of Building Intelligence Group and a consultant with more than twenty five years of experience in a variety of industries including information technology, telecommunications, and financial services. He holds a BS in Computer Engineering and Masters in Computing and Information Sciences from Case Western Reserve University and an MBA from University of San Francisco.

Visit www.Intelligent-Building-Dictionary.com for:

- Fast, easy, online lookup with exact and closest matches.

- Hyperlinks to information resources.

- Additional terms, updates, and errata.

- Information on new editions and other formats.

Give us your feedback online:

- Tell us what you like about this dictionary.

- Let us know how we can make it more useful.

See www.HandsOnGuide.com for:

- Other great Intelligent Building titles.

- Quantity discounts on bulk purchases.

Want a customized Dictionary for your membership organization or marketing program? We can do that, contact sales@handsonguide.com for details.

Interested in writing for the Intelligent Building market? Hands-on-Guide is looking for authors, send your proposal to sales@handsonguide.com.

Printed in the United States
109492LV00004B/348/A

Fig. 16

TIA-942 Data Center Cabling

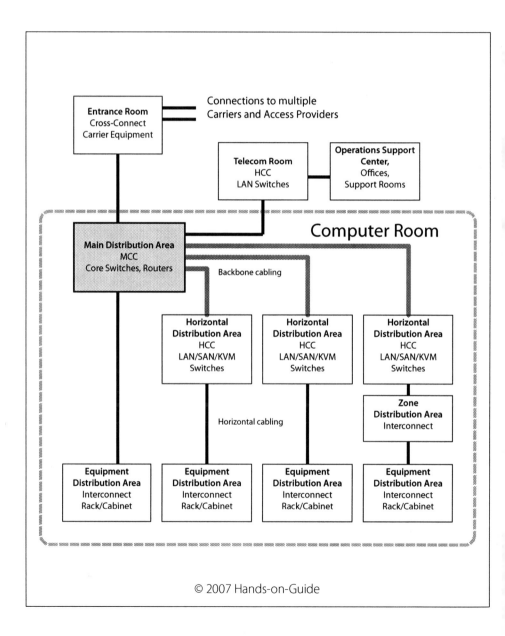